Progress in Nonlinear Differential Equations and Their Applications
Volume 30

Pablo Pedregal

Parametrized Measures and Variational Principles

Birkhäuser
Basel · Boston · Berlin

Pablo Pedregal
E.T.S.I. Industriales
Universidad de Castilla-La Mancha
13071 Ciudad Real
Spain

1991 Mathematics Subject Classification: 49-02, 49J40, 49J45, 49M20

A CIP catalogue record for this book is available from the Library of Congress, Washington D.C., USA

Deutsche Bibliothek Cataloging-in-Publication Data
Pedregal, Pablo:
Parametrized measures and variational principles / Pablo Pedregal. –
Basel ; Boston ; Berlin : Birkhäuser, 1997
(Progress in nonlinear differential equations and their applications ;
Vol. 30)
ISBN 3-7643-5697-9 (Basel ...)
ISBN 0-8176-5697-9 (Boston)

Printed on acid-free paper produced from chlorine-free pulp. TCF ∞
Printed in Germany
ISBN 3-7643-5697-9
ISBN 0-8176-5697-9

9 8 7 6 5 4 3 2 1

Table of Contents

To my wife Josefa

and

to our children Daniel and Silvia;

because of them all,

mathematics makes so much sense to me.

Preface

Weak convergence is a basic tool of modern nonlinear analysis because it enjoys the same compactness properties that finite dimensional spaces do: basically, bounded sequences are weak relatively compact sets. Nonetheless, weak convergence does not behave as one would desire with respect to nonlinear functionals and operations. This difficulty is what makes nonlinear analysis much harder than would normally be expected. Parametrized measures is a device to understand weak convergence and its behavior with respect to nonlinear functionals. Under suitable hypotheses, it yields a way of representing through integrals weak limits of compositions with nonlinear functions. It is particularly helpful in comprehending oscillatory phenomena and in keeping track of how oscillations change when a nonlinear functional is applied.

Weak convergence also plays a fundamental role in the modern treatment of the calculus of variations, again because uniform bounds in norm for sequences allow to have weak convergent subsequences. In order to achieve the existence of minimizers for a particular functional, the property of weak lower semicontinuity should be established first. This is the crucial and most delicate step in the so-called direct method of the calculus of variations. A fairly large amount of work has been devoted to determine under what assumptions we can have this lower semicontinuity with respect to weak topologies for nonlinear functionals in the form of integrals. The conclusion of all this work is that some type of convexity, understood in a broader sense, is usually involved. Another fundamental question is how to deal with functionals that do not enjoy this weak lower semicontinuity property. In some cases existence of minimizers can still be attained, but in many instances the lack of convexity leads to oscillating minimizing sequences whose weak limit is not a minimizer. Precisely, in the context of these irregular variational problems, parametrized measures were introduced by Young to understand and predict the highly oscillatory nature of minimizing sequences. Recently, they have been successfully used in some problems in partial differential equations and in the analysis of solid-solid phase transitions in the context of nonlinear elasticity. These developments have led to a renewed interest in parametrized measures.

Our contribution consists in the realization that, because parametrized measures serve to represent weak limits of nonlinear functionals, they can also be used to understand weak lower semicontinuity. As a matter of fact, almost any weak lower semicontinuity result (that involves an inequality) can be recast, from this standpoint, as a particular Jensen's inequality with respect to the parametrized measure. The connection is even deeper: Jensen's inequality provides the characterization of particular families of parametrized measures that can be generated by sequences of functions with special properties. We show in this fashion that having the characterization of parametrized measures is all we need to understand weak lower semicontinuity and relaxation as well. These characterizations are also important for nonconvex problems because in such cases a generalized variational principle may be set up where parametrized measures are admissible to enter the minimization process. The conditions on admissibility restrict the families of probability measures that can compete in a particular variational problem. Understanding these constraints is a difficult task in some cases. In the context of gradient parametrized measures this was the original motivation that pushed us to look for characterizations of parametrized measures that can be generated by sequences of gradients. From this point of view, weak convergence reduces to manipulating families of probability measures. The crucial issue is to comprehend how restrictions on these families of measures provide properties on generating, weak convergent sequences. For all these reasons, we place the emphasis on characterizing particular families of parametrized measures and then deal rather easily with the two basic questions in the calculus of variations: weak lower semicontinuity and nonconvexity.

This program has been carried out over the years in different papers. Most of the results in Chapters 6, 7, 8 and 10 can be found in [186], [187], [188], [189], [190], [191], [257], [259], [260]. In these chapters we concentrate on the variational case where we are interested in analyzing parametrized measures that can be generated by sequences of gradients or sequences of functions whose divergences make up a compact set of distributions. Some other related situations have been studied in [142], [192], [255], [258]. The first chapter introduces parametrized measures and our point of view based on characterizations in a nontechnical fashion. The goal is to motivate and present an overall picture of the subject. Precise statements, however, are needed when discussing the curl constraint in order to apply results with rigor in Chapters 3, 4 and 5. These three chapters contain the main applications (weak lower semicontinuity and relaxation) from our perspective on the subject. Chapter 2 gathers a number of interesting (although simpler) situations in which parametrized measures help in seeking answers to problems. It is a sort of intermediate step for the ideas ahead. Finally, Chapter 9 explores the relationship of quasiconvexity and rank-one convexity and ends up with the remarkable counterexample of V. Sverak.

To summarize we can say that by means of parametrized measures a unified approach to the fundamental questions of the calculus of variations can be developed. This has been the motivation and the goal of this book.

The author has tried to build a self-contained text. Some material that is not standard has been collected in a series of appendices at the end of selected chapters. Some familiarity is assumed with many topics on functional analysis, measure theory, Sobolev spaces, etc. A partial list of such topics include duality, reflexivity, weak and weak * topologies, the Banach-Alaouglu-Bourbaki theorem, the Hahn-Banach theorem, distributions and test functions, L^p-spaces and Sobolev spaces, equiintegrability and failure of equiintegrability in L^1, the compactness theorem for Sobolev spaces, Poincaré's inequality, the Lax-Milgram lemma, local Sobolev spaces, Hölder's inequality for integrals and for series, maximal functions and operators, Radon measures, duality of continuous functions and measures, total variation, the monotone convergence and dominated convergence theorems, the Radon-Nykodym theorem, the Lebesgue differentiation theorem, Fatou's lemma. Some knowledge of convex analysis, upper semicontinuity, one-point compactifications and the mean-value theorem for integrals is also assumed. References are given where many such topics can be studied or reviewed if necessary.

Many people have contributed in some way or another to this enterprise. First of all, I want to thank most sincerely David Kinderlehrer. I can certainly say that this text would have never been more than a collection of disconnected ideas without his encouragement, support, enthusiasm and helpful comments. It is not necessary to stress the fact that he enriched and shared many of the key ideas the reader will encounter in this book. My thanks go also to Irene Fonseca and Stefan Müller. Some shorter, more elegant proofs of delicate technical points in Chapter 8 are due to them. I also appreciate the help of Richard Jordan concerning the section on turbulent fluids. I learnt from him how parametrized measures can help in the description of long time behavior of complicated systems. I must also show my gratitude to J. Ball, M. Chipot, J. Ericksen, M. Luskin, R. James, R. Kohn, V. Sverak and L. Tartar. In lectures and discussions I learnt from them to appreciate and to deepen my knowledge of variational principles, nonlinear elasticity, computational issues, weak convergence, quasiconvexity, etc.

Finally, I want to thank the staff of the Editorial Department of Birkhäuser for their efficiency and their interest in improving the final version of this text.

Ciudad Real, Spain

Pablo Pedregal
September, 1996

Chapter 1
Introduction

1. The direct method

The historical problem of the calculus of variations and of the theory of optimization is that of finding minimizers of functionals in the form of integrals defined over infinite dimensional spaces. Historically, these problems were tackled and in many instances solved through the associated Euler-Lagrange equation, which is the analogue of the critical point condition for functions defined over finite dimensional spaces. This condition usually leads to an equation or system of ordinary differential equations or partial differential equations. The search for minimizers (or in general extremals) was reduced in this way to finding certain solutions of differential equations associated to the corresponding functional. Whenever these solutions could be found explicitly or shown to exist, one would establish, under suitable assumptions, the existence of solutions to the variational principle. This way of proceeding is especially fruitful in one dimension, when the Euler-Lagrange equation is an ordinary differential equation or system. In higher dimensions nonlinear partial differential equations need to be solved and in general it is not an easy task to show existence of solutions. Consequently, attention was focussed on finding extremals directly from the functional itself: the direct method of the calculus of variations was the outcome. This method has been so successful that today it is one of the usual ways of showing existence of solutions to many nonlinear elliptic partial differential equations. Another main source of problems in the calculus of variations comes from the mathematical theory of nonlinear elasticity, or more generally from continuum mechanics, where experience leads us to postulate that an equilibrium state for some physical system will be a minimizer of some appropriate energy functional pertinent to the model under consideration. We will explain this point of view further in subsequent chapters. We do not pursue here the analysis of the associated Euler-Lagrange equation or system but concentrate on the direct method.

In order to understand the direct method and to isolate the different ingredients we need, let us look for one moment at the finite dimensional situation. Let $I : \mathbf{R}^p \to \mathbf{R}^*$ where we take $\mathbf{R}^* = \mathbf{R} \cup \{+\infty\}$ throughout the book. We would like to find $x_0 \in \mathbf{R}^p$ such that $I(x_0) \leq I(x)$ for all $x \in \mathbf{R}^p$. First of all we need to make sure that I is bounded from below, $I(x) \geq c > -\infty$, for all $x \in \mathbf{R}^p$. Otherwise, there is nothing we can do about the analysis of the minimization problem. Let

$$-\infty < m = \inf \{I(x) : x \in \mathbf{R}^p\},$$

and let $\{x_n\}$ be a minimizing sequence: $I(x_n) \to m$. If $\{x_n\}$ is relatively compact in $\mathbf{R}^p$ (this is the case if $\liminf_{x\to\infty} I(x) > m$) and I is continuous, for some suitable subsequence, not relabeled, $x_n \to x_0$ and $I(x_n) \to m$. Therefore $I(x_0) = m$ and x_0 is a minimizer. In fact, if we are interested in minimizers it is enough to ask for the lower semicontinuity of I:

$$I(x) \leq \liminf_{n\to\infty} I(x_n),$$

whenever $x_n \to x$.

The direct method consists in imitating the finite dimensional case, and trying to use it in the infinite dimensional situation. The different important ingredients are:

1. I is not identically $+\infty$;
2. I is bounded from below;
3. Good compactness properties for the topology on the set of competing functions;
4. I should be lower semicontinuous with respect to the chosen topology.

The function spaces of competing functions usually are Banach spaces with integral norms $L^p(\Omega)$, $W^{1,p}(\Omega)$, and the appropriate topologies with good compactness properties are the weak topologies over these spaces. In particular, if X is one of these spaces and is reflexive, it is well known that

$$\|u_n\|_X \leq M < \infty \quad \text{implies} \quad u_n \rightharpoonup u, \; u \in X,$$

possibly for a subsequence (Banach-Alaouglu-Bourbaki Theorem). This property is extremely convenient and explains, from our perspective, why weak convergence is so important and why we are interested in deepening our understanding of it.

Finally, the most difficult step in applying the direct method is to have the lower semicontinuity property with respect to these weak topologies

$$u_n \rightharpoonup u \text{ in } X \quad \text{implies} \quad I(u) \leq \liminf_{n\to\infty} I(u_n).$$

We can summarize the previous considerations in the following abstract theorem.

Theorem 1.1 *Let us consider the variational principle*

$$\min\{I(u) : u \in \mathcal{A}\},$$

where:

i) $\mathcal{A}$ *is a closed, convex subset of a reflexive, Banach space* X*;*
ii) I *is coercive:* $I(u) \geq C\|u\|_X$, $C > 0$, *or* $\lim_{\|u\|\to\infty} I(u) = +\infty$*;*
iii) I *is lower semicontinuous with respect to the weak topology in* X*;*
iv) There exists $\overline{u} \in \mathcal{A}$ *such that* $I(\overline{u}) < \infty$*.*

Then there exists $u_0 \in \mathcal{A}$ *with* $I(u_0) \leq I(u)$ *for all* $u \in \mathcal{A}$*.*

When the functional I is not weak lower semicontinuous (because of some lack of convexity), we talk about relaxation, relaxed functionals and generalized variational principles. We will discuss all of these issues from a general point of view in Section 3. The central topic of this book is the analysis of weak lower semicontinuity and relaxation in different contexts by using parametrized measures. The functional I we will concentrate on, although it is not the only important possibility, is

$$I(u) = \int_\Omega \varphi(x, u(x), \nabla u(x))\, dx$$

where

$$u : \Omega \subset \mathbf{R}^N \to \mathbf{R}^m, \quad \varphi : \Omega \times \mathbf{R}^m \times \mathbf{M}^{m\times N} \to \mathbf{R}^*.$$

2. Parametrized measures: some examples

We study some simple examples to emphasize the behavior of weak convergence with respect to nonlinear functionals. Heuristically, weak convergence means convergence in the average: we say that f_j converges weakly to f, and we write $f_j \rightharpoonup f$, if roughly

$$\int_E f_j\, dx \to \int_E f\, dx,$$

for every subset E. We will be more precise later.

1. Let $\Omega = (0, \pi/2)$ and $f_j(x) = \sin(jx)$, $g_j(x) = f_j^2(x)$, $x \in \Omega$. From elementary trigonometry we have

$$\int_a^b \sin(jx)\, dx = \frac{1}{j}\left(\cos(ja) - \cos(jb)\right),$$

$$\int_a^b \sin^2(jx)\, dx = \frac{(b-a)}{2} + \frac{1}{4j}\left(\sin(2ja) - \sin(2jb)\right).$$

Thus

$$\lim_{j\to\infty} \int_a^b f_j(x)\, dx = 0, \quad \lim_{j\to\infty} \int_a^b g_j(x)\, dx = \frac{(b-a)}{2},$$

for any interval $(a,b) \subset \Omega$. This actually means that

$$f_j \rightharpoonup 0, \qquad g_j \rightharpoonup \frac{1}{2}.$$

Observe that the square of the weak limit does not coincide with the weak limit of the squares. Let us further take $\varphi(x) = \sqrt{x}$ for $x \geq 0$, and let us examine the weak limit of the sequence $\{\varphi(g_j)\}$. In this case, using the periodicity of sin, we get

$$\begin{aligned} \int_a^b |\sin(jx)|\, dx =& \frac{1}{j} \int_{ja}^{jb} |\sin(y)|\, dy \\ =& \frac{2(b-a)}{\pi} + \frac{2}{j}\left(\langle \frac{j(b-a)}{\pi} \rangle - \frac{j(b-a)}{\pi}\right) \\ &+ \frac{1}{j} \int_{ja+\pi\langle j(b-a)/\pi \rangle}^{jb} |\sin(y)|\, dy, \end{aligned}$$

where $\langle a \rangle$ stands for the integer part of a. Due to the two facts

$$\left| \langle \frac{j(b-a)}{\pi} \rangle - \frac{j(b-a)}{\pi} \right| \leq 1,$$
$$j(b-a) - \pi \langle \frac{j(b-a)}{\pi} \rangle \leq \pi,$$

we conclude that the last two terms in the computations above converge to 0, and therefore

$$\varphi(g_j) \rightharpoonup \frac{2}{\pi}.$$

We notice that the weak limit of $\{\varphi(g_j)\}$ is not the composition of φ with the weak limit of $\{g_j\}$. We see that weak convergence does not behave as one would have anticipated with respect to nonlinear functionals.

2. Let $f(x) = 2\chi_{[0,1/2]}(x) - 1$ for $x \in [0,1]$, and extend f to all of $\mathbf{R}$ by periodicity. Take $f_j(x) = f(jx)$. It is easy to see that the jumps from 1 to -1 occur in a smaller and smaller scale as j tends to ∞. On the other hand it is not hard to see that $f_j \rightharpoonup 0$ for $\Omega = (0,1)$. However $f_j^2 \equiv 1$ for all j so that $f_j^2 \rightharpoonup 1$ and once again the square of the weak limit is not the weak limit of the squares. Moreover, if $\varphi : \mathbf{R} \to \mathbf{R}$ is any continuous function then

$$\int_0^1 \varphi(f_j(x))\, dx \to \frac{1}{2}(\varphi(1) + \varphi(-1)),$$

and

$$\int_a^b \varphi(f_j(x))\, dx \to \frac{(b-a)}{2}(\varphi(1) + \varphi(-1))$$

for any interval $(a, b) \subset (0, 1)$; hence for any such φ

$$\varphi(f_j) \rightharpoonup \frac{1}{2}(\varphi(1) + \varphi(-1)).$$

3. Consider the three sequences of functions

$$f_j, g_j, h_j : \Omega \subset \mathbf{R}^2 \to \mathbf{R},$$

defined by

$$\begin{aligned} f_j(x, y) &= \sin(jy), \\ g_j(x, y) &= \cos(jx), \\ h_j(x, y) &= \sin(j(x - y)). \end{aligned}$$

As in Example 1, it is not hard to see that the weak limit for the three sequences is 0,

$$f_j, g_j, h_j \rightharpoonup 0.$$

Let $F : \mathbf{R}^3 \to \mathbf{R}$ be defined by $F(u, v, w) = uvw$. In this case and using basic trigonometric formulas

$$\begin{aligned} F(f_j, g_j, h_j) &= \sin(jy)\cos(jx)\sin(j(x - y)) \\ &= \sin(jy)\cos(jx)\sin(jx)\cos(jy) - \sin^2(jy)\cos^2(jx) \\ &= \frac{1}{4}\sin(2jy)\sin(2jx) - \sin^2(jy)\cos^2(jx). \end{aligned}$$

Since in the last equality the variables are separated, it is easy to obtain, integrating in a rectangle $[a, b] \times [c, d]$ and bearing in mind the computations in Example 1, that

$$F(f_j, g_j, h_j) \rightharpoonup -\frac{1}{4}.$$

However, $F(0, 0, 0) = 0$. It is interesting to notice that in this case we have the reverse inequality $-1/4 < 0$. This example is a classical counterexample in the context of compensated compactness. It is also relevant to the problem of showing that rank-one convexity does not imply quasiconvexity (Chapter 9).

Weak convergence is particularly well suited to deal with bounded sequences in $L^\infty(\Omega)$. We say that a sequence $\{f_j\}$ in $L^\infty(\Omega)$ converges weakly * to f (in $L^\infty(\Omega)$) and is denoted by $f_j \overset{*}{\rightharpoonup} f$ in $L^\infty(\Omega)$, if

$$\int_\Omega g f_j \, dx \to \int_\Omega g f \, dx,$$

for all g in $L^1(\Omega)$. The reason for the * is well known from functional analysis and is due to the lack of reflexivity of $L^1(\Omega)$. We can always extract a weak * convergent subsequence from a bounded sequence in $L^\infty(\Omega)$ (Banach-Alaouglu-Bourbaki Theorem).

The examples examined above show that there is something special about sequences of a highly oscillatory nature. In general terms the problem we would like to address is the following. Suppose that $f_j \overset{*}{\rightharpoonup} f$ in $L^\infty(\Omega)$ so that $\|f_j\|_{L^\infty(\Omega)} \le C < \infty$, and φ is a continuous function. $\{\varphi(f_j)\}$ is also a uniformly bounded sequence in $L^\infty(\Omega)$. Hence, a subsequence converges weakly * in $L^\infty(\Omega)$

$$\varphi(f_j) \overset{*}{\rightharpoonup} g \quad \text{in } L^\infty(\Omega).$$

The question is: What is g? The answer is not $\varphi(f)$ as we have tried to explain through the examples. Parametrized measures is a device to answer this basic question.

A parametrized measure is a family of probability measures $\nu = \{\nu_x\}_{x\in\Omega}$ associated with a sequence of functions $f_j : \Omega \subset \mathbf{R}^N \to \mathbf{R}^m$ such that supp $(\nu_x) \subset \mathbf{R}^m$ and they depend measurably on $x \in \Omega$, which means that for any continuous $\varphi : \mathbf{R}^m \to \mathbf{R}$, the function

$$\overline{\varphi}(x) = \int_{\mathbf{R}^m} \varphi(\lambda)\, d\nu_x(\lambda) = \langle \varphi, \nu_x \rangle \tag{1–1}$$

is measurable. The **fundamental property** of this family of probability measures is that whenever $\{\varphi(f_j)\}$ converges weakly * in $L^\infty(\Omega)$ (or more generally weak in some $L^p(\Omega)$) the weak limit can be identified with the function $\overline{\varphi}$ in (1–1):

$$\lim_{j\to\infty} \int_\Omega \varphi(f_j) g(x)\, dx = \int_\Omega g(x) \int_{\mathbf{R}^m} \varphi(\lambda)\, d\nu_x(\lambda)\, dx, \tag{1–2}$$

for all $g \in L^1(\Omega)$. Intuitively, the parametrized measure can be thought of as giving the limiting probability distribution of the values of $\{f_j\}$ when points are taken randomly around each $x \in \Omega$. Another way of expressing this same idea is the following. Let $f_j : \Omega \subset \mathbf{R}^N \to \mathbf{R}^m$ and let $B_R(x)$ denote the ball of radius $R > 0$ centered at $x \in \Omega$. For any measurable subset $E \subset \mathbf{R}^m$

$$\nu_x(E) = \lim_{R\to 0} \lim_{j\to\infty} \frac{|\{y \in B_R(x) : f_j(y) \in E\}|}{|B_R(x)|}, \tag{1–3}$$

where bars $|\cdot|$ denote Lebesgue measure. One important question to be addressed (Chapter 6) is why these limits exist and why they define Radon measures ν_x. The connection between (1–2) and (1–3) is established by taking g in (1–2) to localize around $x \in \Omega$ ($g \equiv \delta_x$, δ the Dirac mass) and φ close to the characteristic function of E, χ_E.

Let us look at some other examples.

4. Let $g : \mathbf{R}^2 \to \mathbf{R}$ be defined by

$$g(x_1, x_2) = \int_0^{x_1+x_2} \chi_{3/4}(s)\, ds,$$

where $\chi_{3/4}$ is the characteristic function of the interval $(0,3/4) \subset (0,1)$ extended periodically to all of $\mathbf{R}$. Let $\overline{u} : \mathbf{R}^2 \to \mathbf{R}^2$ be defined by $\overline{u}(x) = (g(x), g(x))$, $x = (x_1, x_2) \in \mathbf{R}^2$ and $u_j(x) = (1/j)\overline{u}(jx)$. If we compute the gradients of these functions we obtain

$$\nabla u_j(x) = \nabla \overline{u}(jx) = \chi_{3/4}(j(x_1 + x_2))\,(1,1) \otimes (1,1),$$

or in matrix form

$$\nabla u_j(x) = \chi_{3/4}(j(x_1 + x_2)) \begin{pmatrix} 1 & 1 \\ 1 & 1 \end{pmatrix}$$
$$= \begin{cases} \begin{pmatrix} 1 & 1 \\ 1 & 1 \end{pmatrix}, & 0 < j(x_1 + x_2) - \langle j(x_1 + x_2)\rangle < \frac{3}{4}, \\ \begin{pmatrix} 0 & 0 \\ 0 & 0 \end{pmatrix}, & \frac{3}{4} < j(x_1 + x_2) - \langle j(x_1 + x_2)\rangle < 1. \end{cases}$$

Here, the tensor product $a \otimes n$ is another way of writing the rank-one matrix $a\,n^T$ for $a \in \mathbf{R}^m$ and $n \in \mathbf{R}^N$. We are interested in determining the parametrized measure associated to this sequence of gradients. Let

$$A = \begin{pmatrix} 1 & 1 \\ 1 & 1 \end{pmatrix}, \quad O = \begin{pmatrix} 0 & 0 \\ 0 & 0 \end{pmatrix}.$$

Certainly, if $E \subset \mathbf{M}^{2\times 2}$ contains neither A nor O then

$$\lim_{R\to 0}\lim_{j\to\infty} \frac{|\{y \in B_R(x) : \nabla u_j(y) \in E\}|}{|B_R(x)|} = 0,$$

which means according to (1–3) (in this case $f_j = \nabla u_j$) that ν_x is concentrated in A and O, and so

$$\nu_x = \lambda(x)\delta_A + (1 - \lambda(x))\,\delta_O, \quad \lambda(x) \in [0,1].$$

Moreover, if one thinks about how A and O are distributed by ∇u_j,

$$\lambda(x) = \lim_{R\to 0}\lim_{j\to\infty} \frac{|\{y \in B_R(x) : \nabla u_j(y) = A\}|}{|B_R(x)|} = \frac{3}{4},$$

independently of $x \in \Omega$. Hence

$$\nu = \frac{3}{4}\delta_A + \frac{1}{4}\delta_O.$$

For any continuous function $\varphi : \mathbf{M}^{2\times 2} \to \mathbf{R}$ we have

$$\varphi(\nabla u_j) \overset{*}{\rightharpoonup} \frac{3}{4}\varphi(A) + \frac{1}{4}\varphi(O) = \langle \varphi, \nu\rangle \quad \text{in } L^\infty(\Omega). \tag{1–4}$$

5. Let us follow through with the previous example. Take $\varphi_0(F) = |F|^2$ for $F \in \mathbf{M}^{2\times 2}$ so that

$$\begin{aligned}\varphi_0(\nabla u_j) &= |\nabla u_j|^2 \\ &= \begin{cases} 4, & 0 < j(x_1+x_2) - \langle j(x_1+x_2)\rangle < \frac{3}{4}, \\ 0, & \frac{3}{4} < j(x_1+x_2) - \langle j(x_1+x_2)\rangle < 1. \end{cases}\end{aligned}$$

We know that the sequence of gradients $\{\nabla u_j\}$ converges weak * in $L^\infty(\Omega)$ to $\nabla u = (3/4)A + (1/4)O = (3/4)A$ by simply taking φ to be the identity in (1–4). Since φ_0 is convex, by Jensen's inequality (see the Appendix, Section 6 below)

$$\int_{\mathbf{M}^{2\times 2}} |F|^2 \, d\nu(F) \geq \left| \int_{\mathbf{M}^{2\times 2}} F \, d\nu(F) \right|^2,$$

which in this simple case is nothing but

$$\frac{3}{4}\left|\begin{pmatrix} 1 & 1 \\ 1 & 1 \end{pmatrix}\right|^2 + \frac{1}{4}\left|\begin{pmatrix} 0 & 0 \\ 0 & 0 \end{pmatrix}\right|^2 \geq \left|\frac{3}{4}\begin{pmatrix} 1 & 1 \\ 1 & 1 \end{pmatrix} + \frac{1}{4}\begin{pmatrix} 0 & 0 \\ 0 & 0 \end{pmatrix}\right|^2.$$

By (1–4) applied to φ_0, this inequality is exactly

$$\lim_{j\to\infty} \int_\Omega |\nabla u_j|^2 \, dx \geq \int_\Omega |\nabla u|^2 \, dx,$$

a weak lower semicontinuity result.

3. A general framework

We describe in this section how parametrized measures can be used to understand the basic questions of the calculus of variations, in particular weak lower semicontinuity and relaxation. We have already pointed out heuristically what a parametrized measure is and how it serves to represent weak limits of compositions with the specific sequence it comes from.

Assume that for some reason we are interested in a particular class of sequences of $L^\infty(\Omega)$-functions taking values in $\mathbf{R}^m$, uniformly bounded in the $L^\infty(\Omega)$-norm, and let us denote such class by $\mathcal{L}$. We may think of $\mathcal{L}$ as incorporating all the constraints or restrictions for the sequences we would like to consider in some particular situation of interest. Typical examples are

$$\begin{aligned}
&\mathcal{L} = \{\{u_j\} \text{ bounded in } L^\infty(\Omega)\}, \quad \text{(no differential constraint)}, \\
&\mathcal{L} = \{\{u_j\} : \operatorname{curl} u_j = 0\}, \quad \text{(the gradient case)}, \\
&\mathcal{L} = \left\{\{\nabla v_j\} : v_j - v_0 \in W_0^{1,\infty}(\Omega)\right\}, \quad \text{(Dirichlet boundary condition)}, \\
&\mathcal{L} = \left\{\{u_j\} : \{\operatorname{div} u_j\} \text{ is weakly relatively compact in } H^{-1}(\Omega)\right\}.
\end{aligned}$$

Let $\overline{\mathcal{L}}$ be the set of *homogeneous* parametrized measures associated to sequences in $\mathcal{L}$. What we mean by the word homogeneous is that the family of probability measures $\{\nu_x\}_{x\in\Omega}$ reduces to a unique single measure: $\nu_x = \nu$ for a.e. $x \in \Omega$. $\overline{\mathcal{L}}$ is a subset of probability measures supported on $\mathbf{R}^m$ and for each one of these, ν, there exists a sequence $\{f_j\}$ in $\mathcal{L}$ such that (1–2) holds when $\nu_x = \nu$ for a.e. $x \in \Omega$ and any continuous φ. For a constant vector $u \in \mathbf{R}^m$ let $\mathcal{A}_u$ stand for the set

$$\mathcal{A}_u = \left\{\nu \in \overline{\mathcal{L}} : u = \int_{\mathbf{R}^m} \lambda \, d\nu(\lambda)\right\}.$$

Given any $\psi : \mathbf{R}^m \to \mathbf{R}^*$, we define the $\mathcal{L}$-convexification of ψ by

$$\begin{aligned}\psi^{\mathcal{L}}(u) &= \inf\left\{\lim_{j\to\infty} \frac{1}{|\Omega|}\int_\Omega \psi(u_j(x))\, dx : \{u_j\} \in \mathcal{L}, u_j \stackrel{*}{\rightharpoonup} u \text{ in } L^\infty(\Omega)\right\}\\ &= \inf\left\{\int_{\mathbf{R}^m} \psi(\lambda)\, d\nu(\lambda) : \nu \in \mathcal{A}_u\right\}.\end{aligned}$$

Notice that $\psi^{\mathcal{L}} \le \psi$ always and possibly $\psi^{\mathcal{L}} = -\infty$. We declare ψ as $\mathcal{L}$-convex if $\psi^{\mathcal{L}} = \psi$, or equivalently if

$$\psi(u) \le \int_{\mathbf{R}^m} \psi(\lambda)\, d\nu(\lambda),$$

for all $\nu \in \mathcal{A}_u$ and all $u \in \mathbf{R}^m$. Notice how this definition is motivated by the weak lower semicontinuity property. The basic explicit assumptions we want to make are

i) *$\psi^{\mathcal{L}}$ is well defined independently of the choice of Ω;*

*ii) $\mathcal{A}_u$ is convex and closed with respect to the weak * topology in the space of finite Radon measures over $\mathbf{R}^m$;*

iii) localization: if $\nu = \{\nu_x\}_{x\in\Omega}$ can be generated by a sequence in $\mathcal{L}$ then for a.e. $x \in \Omega$, $\nu_x \in \mathcal{A}_{u(x)}$ where $u(x)$ is the first moment of ν_x: $u(x) = \int \lambda\, d\nu_x(\lambda)$.

Under these important hypotheses and some other technicalities that depend on the particular $\mathcal{L}$ and the fact that we would like to work with exponent p finite, one can obtain the following characterization of those parametrized measures that can be generated by sequences in $\mathcal{L}$: $\nu = \{\nu_x\}_{x\in\Omega}$, an arbitrary family of probability measures, can be generated by a sequence in $\mathcal{L}$ (by this we mean that there exists a sequence $\{u_j\} \in \mathcal{L}$ for which (1–2) is valid) if and only if

$$\varphi\left(\int_{\mathbf{R}^m} \lambda\, d\nu_x(\lambda)\right) \le \int_{\mathbf{R}^m} \varphi(\lambda)\, d\nu_x(\lambda) \tag{1–5}$$

for a.e. $x \in \Omega$ and every continuous, $\mathcal{L}$-convex function φ. As pointed out, there are also some other compatibility conditions and technicalities involved that are not as important as (1–5). Based on this characterization in terms of Jensen's inequality (1–5), one can prove rather easily weak lower semicontinuity facts and relaxation theorems.

Suppose we are in the situation

$$u_j \overset{*}{\rightharpoonup} u \quad \text{in } L^\infty(\Omega), \qquad \{u_j\} \in \mathcal{L}.$$

We would like to understand the restrictions on φ that ensure weak lower semicontinuity for the functional

$$\int_\Omega \varphi(u(x))\,dx,$$

that is to say

$$\int_E \varphi(u)\,dx \le \liminf_{j\to\infty} \int_E \varphi(u_j)\,dx, \quad E \subset \Omega.$$

The necessary and sufficient condition is that φ be $\mathcal{L}$-convex. Let us sketch the proof. For any constant vector $u \in \mathbf{R}^m$, using the weak lower semicontinuity,

$$\begin{aligned}\varphi^{\mathcal{L}}(u) &= \inf\left\{\liminf_{j\to\infty} \frac{1}{|\Omega|}\int_\Omega \varphi(u_j)\,dx : u_j \overset{*}{\rightharpoonup} u, \{u_j\} \in \mathcal{L}\right\}\\ &\ge \varphi(u)\\ &\ge \varphi^{\mathcal{L}}(u),\end{aligned}$$

so that φ is $\mathcal{L}$-convex. Conversely, if φ is $\mathcal{L}$-convex then Jensen's inequality (1–5) holds for $\nu = \{\nu_x\}_{x\in\Omega}$, the parametrized measure generated by $\{u_j\}$ (or some appropriate subsequence), and therefore

$$\begin{aligned}\lim_{j\to\infty} \int_E \varphi(u_j)\,dx &= \int_E \int_{\mathbf{R}^m} \varphi(\lambda)\,d\nu_x(\lambda)\,dx\\ &\ge \int_E \varphi(\int_{\mathbf{R}^m} \lambda\,d\nu_x(\lambda))\,dx\\ &= \int_E \varphi(u)\,dx.\end{aligned}$$

Once this lower semicontinuity is accomplished, the direct method of the calculus of variations provides existence of minimizers under coercivity assumptions.

Concerning relaxation we have to consider non-lower semicontinuous functionals

$$I(u) = \int_\Omega \psi(u)\,dx, \quad u \in \mathcal{C},$$

where ψ is not $\mathcal{L}$-convex and the set of admissible functions u is $\mathcal{C}$. In this situation we take $\mathcal{L} = \{\{u_j\} \text{ bounded in } L^\infty(\Omega) : u_j \in \mathcal{C}\}$. The so-called relaxed functional is

$$I^{\mathcal{L}}(u) = \int_\Omega \psi^{\mathcal{L}}(u)\,dx, \quad u \in \mathcal{C}.$$

A typical relaxation result is

$$\inf\{I(u) : u \in \mathcal{C}\} = \min\left\{I^{\mathcal{L}}(u) : u \in \mathcal{C}\right\}.$$

Again the proof is not difficult if we rely on the characterization in terms of Jensen's inequality. Let m and $m^{\mathcal{L}}$ be the two infima. Since $\psi^{\mathcal{L}} \leq \psi$, $m^{\mathcal{L}} \leq m$. For $u \in \mathcal{C}$,

$$\psi^{\mathcal{L}}(u(x)) = \inf\left\{\int_{\mathbf{R}^m} \psi(\lambda)\, d\nu(\lambda) : \nu \in \mathcal{A}_{u(x)}\right\}.$$

Assume we can count on some other hypothesis which enables us to assert that *the last infimum is in fact a minimum* so that the infimum is attained

$$\psi^{\mathcal{L}}(u(x)) = \int_{\mathbf{R}^m} \psi(\lambda)\, d\nu_x(\lambda), \quad \nu_x \in \mathcal{A}_{u(x)}.$$

Consider now the family of probability measures $\nu = \{\nu_x\}_{x\in\Omega}$. For a $\mathcal{L}$-convex function φ, by definition,

$$\varphi(u(x)) = \varphi\left(\int_{\mathbf{R}^m} \lambda\, d\nu_x(\lambda)\right) \leq \int_{\mathbf{R}^m} \varphi(\lambda)\, d\nu_x(\lambda).$$

The conclusion is that ν can be generated by $\mathcal{L}$, according to the characterization in terms of Jensen's inequality (1–5): there exists a sequence, $\{u_j\}$, in $\mathcal{L}$ such that by construction

$$\int_\Omega \psi(u_j)\, dx \to \int_\Omega \int_{\mathbf{R}^m} \psi(\lambda)\, d\nu_x(\lambda)\, dx = \int_\Omega \psi^{\mathcal{L}}(u)\, dx.$$

This implies that $m = m^{\mathcal{L}}$.

To explain from a different viewpoint why it is important to have characterizations of parametrized measures, let us examine further the variational principle

$$I(u) = \int_\Omega \psi(u)\, dx, \quad u \in \mathcal{C}, \tag{1–6}$$

where $\mathcal{C}$ includes all of the admissibility constraints and once again we set

$$\mathcal{L} = \{\{u_j\} \text{ bounded in } L^\infty(\Omega) : u_j \in \mathcal{C}\}.$$

Let us postulate that ψ is not $\mathcal{L}$-convex so that, in general, minimizers might not exist. In this case we can set up a generalized variational problem

$$\tilde{I}(\nu) = \int_\Omega \int_{\mathbf{R}^m} \psi(\lambda)\, d\nu_x(\lambda)\, dx, \tag{1–7}$$

where $\nu = \{\nu_x\}_{x\in\Omega}$ is a family of probability measures. Obviously, if $\nu_x = \delta_{u(x)}$ for $x \in \Omega$ and $u \in \mathcal{C}$ then $I(u) = \tilde{I}(\nu)$ so that the information coming from the original functional is somehow included in $\tilde{I}$. If we want $\tilde{I}$ to be relevant to I, which is our goal, the admissibility conditions for ν to compete in the minimization problem (1–7) should incorporate the fact that ν must be generated as a parametrized measure from sequences of $\mathcal{L}$. Otherwise, the minimization problem (1–7) might not have any significance with respect to (1–6). If we can understand the restrictions placed on ν so that it can be generated by sequences in $\mathcal{L}$, then those conditions should be included in the admissibility criterion for ν in (1–7). Therefore it is important to be able to provide some way of characterizing the families that can enter into the variational problem for $\tilde{I}$. Once this is done, it is not difficult to show the following two nice facts: (1–7) always admits a solution ν_0, and $\inf I(u) = \tilde{I}(\nu_0)$. The sequences from $\mathcal{L}$ that generate ν_0 are minimizing sequences for I and through them one can understand the behavior of I. The functional $\tilde{I}$ is also closely related to $I^{\mathcal{L}}$, when the appropriate admissibility conditions are considered, as we will see.

One of the most important examples where this approach has been sucessful in providing some (partial) answers to specific questions is the analysis of phase transitions in crystalline solids. Chapter 5 is devoted to this topic. Another situation of interest where generalized variational principles for parametrized measures are important relates to the failure of constraints preserved by weak convergence (see hypothesis ii) above) regardless of whether the functional itself is weak lower semicontinuous or not. In these cases, parametrized measures are needed to provide some description of the behavior of minimizing sequences. An important example in the theory of ferromagnetism is presented in Chapter 5. The localization property iii) might also fail in some cases. Even so parametrized measures are important to set up equivalent generalized variational principles. Parametrized measures are a convenient tool to understand and manipulate weak convergence. They are especially well-suited for variational problems.

The general approach we propose can be summarized in several steps. First, we need to determine the properties of the sequences we are interested in in order to define the class $\mathcal{L}$ (for instance bounded sequences in $W^{1,p}(\Omega)$); secondly, the notion of convexity associated to such sequences ($\mathcal{L}$-convexity) must be clarified and studied; then a result characterizing parametrized measures associated to such sequences in terms of Jensen's inequality is needed; finally, existence and relaxation theorems are established.

In this book we will concentrate on the important case mentioned above where we take $\mathcal{L}$ to be the class of sequences of gradients of functions in $W^{1,p}(\Omega)$ for finite p. This case is, once again, motivated by the functional

$$I(u) = \int_\Omega \varphi(x, u(x), \nabla u(x))\, dx.$$

Under coercivity assumptions on φ with respect to the gradient variable minimizing sequences for I belong to $\mathcal{L}$. Simpler cases will also be examined in Chapter 2. The remaining sections in this chapter are intended to provide a global overview of the last five chapters sufficient to understand Chapters 2, 3, 4 and 5. The emphasis here is placed on the ideas and results while technical proofs are deferred. The main technical difficulties in the analysis contained in those chapters come precisely from the differential constraints (curl or div) and the fact that we want to work in $W^{1,p}(\Omega)$ for *finite* p. In this case, weak convergence for compositions $\{\varphi(\nabla u_j)\}$ where φ grows no faster than the pth power and $\{u_j\}$ is bounded in $W^{1,p}(\Omega)$ is not guaranteed, while this is always true for $p = \infty$ and any continuous φ.

The framework described above is related to a general approach to weak convergence known as "compensated compactness" in which non-trivial weak convergence facts are derived exploiting additional information on the derivatives of the sequence under consideration. The information on the derivatives usually comes in the form of some boundedness properties for a differential operator A with constant coefficients a_{ilk},

$$(Au)_i = \sum_{l,k} a_{ilk} \frac{\partial u^l}{\partial x_k}, \quad i = 1, \dots, s.$$

We put, however, the emphasis on deriving weak convergence facts through a close analysis of the underlying parametrized measure since we are especially interested in variational principles and we do not require more detailed information about oscillations at this stage. The particular cases considered above are $A =$ curl and $A =$ div,

$$\operatorname{curl} u = \frac{\partial u^i_l}{\partial x_k} - \frac{\partial u^i_k}{\partial x_l}, \quad i = 1, \dots, m, \quad l, k = 1, \dots, N,$$

$$\operatorname{div} u = \sum_{k=1}^{N} \frac{\partial u^i_k}{\partial x_k}, \quad i = 1, \dots, m.$$

4. The case curl: gradient parametrized measures

This section is essentially a summary as well as an overview of Chapter 8 and a discussion of the main facts concerning characterizations of parametrized measures under this main differential constraint. At the same time we include the main results to understand our approach to weak lower semicontinuity and relaxation in Chapters 3 and 4. For the complete proofs, we refer the reader to Chapters 7 and 8.

Let us consider vector functions $u : \Omega \subset \mathbf{R}^N \to \mathbf{R}^m$ in $W^{1,\infty}(\Omega)$. The gradient ∇u is a matrix in $\mathbf{M}^{m \times N}$ for every $x \in \Omega$. We would like to analyze the $\mathcal{L}$-convexification within the framework described in the previous section for

$$\mathcal{L} = \left\{\{\nabla u_j\} : \{u_j\} \text{ bounded in } W^{1,\infty}(\Omega)\right\}.$$

For $\psi : \mathbf{M}^{m\times N} \to \mathbf{R}^*$, the associated convexification at Y, $\psi^{\mathcal{L}}(Y)$, is defined through the infimum

$$\inf\left\{\liminf_{j\to\infty} \frac{1}{|\Omega|}\int_\Omega \psi(\nabla u_j(x))\,dx : \{\nabla u_j\} \in \mathcal{L}, \nabla u_j \overset{*}{\rightharpoonup} Y \text{ in } W^{1,\infty}(\Omega)\right\}. \tag{1–8}$$

We want to find a simpler, more direct definition of this convexification.

Let $u_Y(x)$ be the affine, Lipschitz function $u_Y(x) = Yx$ for $x \in \Omega$. For $u \in W^{1,\infty}(\Omega)$ such that $u - u_Y \in W_0^{1,\infty}(\Omega)$ we can produce an admissible sequence in (1–8). This can be done through the Riemann-Lebesgue lemma which is a particular case of a more general homogenization theorem.

Lemma 8.2 *Let Ω be a domain in $\mathbf{R}^N$ (open, bounded, regular subset) and $u \in W^{1,\infty}(\Omega)$, $u - u_Y \in W_0^{1,\infty}(\Omega)$. There exists a sequence $\{u_j\}$ bounded in $W^{1,\infty}(\Omega)$, $u_j - u_Y \in W_0^{1,\infty}(\Omega)$, such that the parametrized measure associated to $\{\nabla u_j\}$ is homogeneous and defined by*

$$\langle \overline{\nu}, \varphi\rangle = \frac{1}{|\Omega|}\int_\Omega \varphi(\nabla u(x))\,dx,$$

for any continuous φ.

In general a bar over a parametrized measure means its average. As an immediate consequence of this fact, we certainly have

$$\psi^{\mathcal{L}}(Y) \le \inf\left\{\frac{1}{|\Omega|}\int_\Omega \psi(\nabla u(x))\,dx : u - u_Y \in W_0^{1,\infty}(\Omega)\right\}. \tag{1–9}$$

Let us denote by $Q\psi(Y)$ the infimum on the right-hand side of (1–9). The Q stands for quasiconvexity because $\mathcal{L}$-convexity for the gradient case is usually called quasiconvexity.

We claim that we have equality in (1–9). Indeed, the following is true: given any bounded sequence in $W^{1,\infty}(\Omega)$, $\{u_j\}$, such that $\nabla u_j \overset{*}{\rightharpoonup} Y$ in $L^\infty(\Omega)$ we can find another bounded sequence in $W^{1,\infty}(\Omega)$, $\{v_j\}$, such that $v_j - u_Y \in W_0^{1,\infty}(\Omega)$ for all j and

$$\lim_{j\to\infty} \frac{1}{|\Omega|}\int_\Omega \psi(\nabla u_j)\,dx = \lim_{j\to\infty} \frac{1}{|\Omega|}\int_\Omega \psi(\nabla v_j)\,dx.$$

This is done by modifying the boundary values of u_j and forcing them to be u_Y. This helpful technique will be used several times in Chapter 4.

Lemma 8.3 *Let $\{v_j\}$ be a bounded sequence in $W^{1,p}(\Omega)$, $p > 1$, such that the sequence $\{\nabla v_j\}$ generates the parametrized measure $\nu = \{\nu_x\}_{x\in\Omega}$. Let*

$$\nabla u(x) = \int_{\mathbf{M}^{m\times N}} A\, d\nu_x(A) \in \mathbf{M}^{m\times N}, \quad u \in W^{1,p}(\Omega),$$

so that $v_j \rightharpoonup u$ in $W^{1,p}(\Omega)$. There exists a new sequence $\{u_k\}$, bounded in $W^{1,p}(\Omega)$, such that $\{\nabla u_k\}$ generates the same parametrized measure ν and $u_k - u \in W_0^{1,p}(\Omega)$ for all k. If for $p < \infty$, $\{|\nabla v_j|^p\}$ is equiintegrable, so is $\{|\nabla u_k|^p\}$.

Therefore we have that

$$Q\psi(Y) = \inf\left\{\frac{1}{|\Omega|}\int_\Omega \psi(Y + \nabla u(x))\, dx : u \in W_0^{1,\infty}(\Omega)\right\} \tag{1–10}$$

is the appropriate convexification for weak lower semicontinuity in the gradient case. A function ψ is called quasiconvex if $Q\psi = \psi$ for all matrices. Equivalently, ψ is quasiconvex (at Y) if

$$\psi(Y) \le \frac{1}{|\Omega|}\int_\Omega \psi(Y + \nabla u(x))\, dx, \tag{1–11}$$

for any $u \in W_0^{1,\infty}(\Omega)$. If ψ is not quasiconvex, its quasiconvexification $Q\psi$ is defined by (1–10). Another important description of the quasiconvexification of a function is given in the next lemma.

Lemma 8.12 *Let $\varphi : \mathbf{M}^{m\times N} \to \mathbf{R}^*$ be an upper-semicontinuous function. The following identity holds*

$$Q\varphi = \sup\{\psi : \psi \le \varphi, \psi \text{ quasiconvex}\}.$$

The quasiconvexity condition (1–11) is hard to understand because we still have an infinite number of test fields $u \in W_0^{1,\infty}(\Omega)$. An important necessary condition can be derived by a basic, fundamental construction. It corresponds to establishing convexity along the directions of the characteristic cone associated to the operator $A = $ curl (the rank-one cone, see Chapter 10 for more details) and consists in looking at a special sequence $\{\nabla u_j\}$ in (1–9). This construction is a generalization of Example 4 in Section 2. Let $Y \in \mathbf{M}^{m\times N}$, $a \in \mathbf{R}^m$ and a unit vector $n \in \mathbf{R}^N$ be given. Let $\chi_{1/2}$ stand for the characteristic function of the interval $(0, 1/2)$ in $(0, 1)$, extended by periodicity to all of $\mathbf{R}$, and $\chi = 2\chi_{1/2} - 1$. Consider the vector function

$$u(x) = Yx + \int_0^{x\cdot n} \chi(s)\, ds\; a, \quad x \in \Omega,$$

and the sequence $u_j(x) = (1/j)u(jx)$, $j \in \mathbf{N}$. Let us examine the gradients

$$\nabla u_j = Y + \nabla u(jx) = Y + \chi(jx \cdot n)\; a \otimes n,$$

(recall that the tensor product $a \otimes n$ is another way of writing the rank-one matrix an^T). We realize that (just as we did in Example 4 in Section 2) the parametrized measure corresponding to $\{\nabla u_j\}$ is

$$\nu = \frac{1}{2}\delta_{Y_1} + \frac{1}{2}\delta_{Y_2},$$

where $Y_1 = Y + a \otimes n$ and $Y_2 = Y - a \otimes n$. In fact, if ψ is any continuous function then

$$\begin{aligned}&\frac{1}{|\Omega|}\int_\Omega \psi(\nabla u_j(x))\,dx - \langle \psi, \nu\rangle \\ &= (|\{x \in \Omega : (jx\cdot n) - \langle jx \cdot n\rangle < 1/2\}| - 1/2)\,\psi(Y_1) \\ &+ (|\{x \in \Omega : (jx\cdot n) - \langle jx \cdot n\rangle > 1/2\}| - 1/2)\,\psi(Y_2),\end{aligned}$$

and the two coefficients for $\psi(Y_1)$ and $\psi(Y_2)$ clearly tend to 0 as $j \to \infty$. In particular, $\nabla u_j \overset{*}{\rightharpoonup} Y$ in $L^\infty(\Omega)$. This last weak convergence can also be proved by the following estimates

$$\left|\frac{1}{j}u(jx) - u_Y(x)\right| = \frac{1}{j}\left|\int_0^{jx\cdot n} \chi(s)\,ds\right|\,|a| \le \frac{|a|}{2j},$$

where we have used the periodicity of χ and the fact that the integral of χ over an integer number of periods vanishes. This uniform convergence means $u_j \rightharpoonup u_Y$ in $W^{1,\infty}(\Omega)$.

The conclusion of these computations is that the sequence $\{u_j\}$ is admissible in (1–8), and hence

$$Q\psi(Y) \le \langle \psi, \nu\rangle = \frac{1}{2}\psi(Y_1) + \frac{1}{2}\psi(Y_2).$$

Observe that, as pointed out above,

$$\int_\Omega \psi(\nabla u_j(x))\,dx \to \langle \psi, \nu\rangle.$$

If ψ is quasiconvex then

$$\psi(Y) \le \frac{1}{2}\psi(Y + a\otimes n) + \frac{1}{2}\psi(Y - a \otimes n),$$

for any matrix Y and vectors $a \in \mathbf{R}^m$ and $n \in \mathbf{R}^N$. This inequality is the rank-one convexity condition which can be reformulated in the following way

$$\psi(Y) \le t\psi(Y_1) + (1-t)\psi(Y_2), \quad t \in [0,1],$$

provided $Y_1 - Y_2 = a \otimes n$ is a rank-one matrix. Rank-one convexity is a necessary condition for quasiconvexity. We will investigate the converse in Chapter 9. In the scalar case $m = 1$ or $N = 1$ all these convexity conditions collapse to the usual convexity because the equality $Y_1 - Y_2 = a \otimes n$ is no longer a restriction.

The basic assumptions i), ii) and iii) described in the setting of the preceding section are correct for the gradient case. Namely, the quasiconvexification (1–10) is well defined regardless of the choice of the domain Ω, the set of homogeneous parametrized measures generated by gradients bounded in $W^{1,\infty}(\Omega)$ is convex and weak * closed and we have a localization principle. The second assertion, although true, is not exactly the conclusion of the second statement below, but it is enough for our purposes. We have the three following facts.

Lemma 8.9 *Let Ω and Ω' be two open, bounded subsets of $\mathbf{R}^N$ with $|\partial\Omega| = |\partial\Omega'| = 0$, and let φ be a function defined on $\mathbf{M}^{m\times N}$. For any matrix Y, the two infima*

$$\frac{1}{|\Omega|} \inf \left\{ \int_\Omega \varphi(\nabla u(x))\, dx : u \in W^{1,\infty}(\Omega), u - u_Y \in W_0^{1,\infty}(\Omega) \right\},$$

and

$$\frac{1}{|\Omega'|} \inf \left\{ \int_{\Omega'} \varphi(\nabla v(x))\, dx : v \in W^{1,\infty}(\Omega'), v - u_Y \in W_0^{1,\infty}(\Omega') \right\}$$

are equal.

Lemma 8.5 *The set*

$$M_Y = \left\{\nu, \text{ probability measure over } \mathbf{M}^{m\times N} : \nu = \overline{\delta_{\nabla u(x)}}, u \in W^{1,\infty}(\Omega),\right.$$
$$\left. u - u_Y \in W_0^{1,\infty}(\Omega)\right\},$$

is convex.

Finally, the localization principle is used to derive necessary conditions. It allows us to understand each individual ν_x as a homogeneous $W^{1,p}$-parametrized measure by itself. We call in general $\nu = \{\nu_x\}_{x\in\Omega}$ a $W^{1,p}$-parametrized measure if it can be generated by a bounded sequence of gradients in $L^p(\Omega)$.

Theorem 8.4 *Let $\{u_j\}$ be a bounded sequence in $W^{1,p}(\Omega)$ and $\nu = \{\nu_x\}_{x\in\Omega}$ the parametrized measure associated to $\{\nabla u_j\}$. Let*

$$F(a) = \int_{\mathbf{M}^{m\times N}} A\, d\nu_a(A)$$

and $u_a(x) = F(a)x$, $a \in \Omega$. For a.e. $a \in \Omega$, there exists a sequence $\{w_j^a\}$ bounded in $W^{1,p}(\Omega)$ such that $w_j^a - u_a \in W_0^{1,p}(\Omega)$, for all j and the parametrized measure associated to $\{\nabla w_j^a\}$ is ν_a, homogeneous.

The main goal of Chapter 8 is the characterization of $W^{1,p}$-parametrized measures, parametrized measures generated by bounded sequences of gradients in $L^p(\Omega)$. We include this result for the finite case $p < \infty$, although the case $p = \infty$ is also correct with minor changes. As pointed out, many of the technicalities with the case $p < \infty$ are related to the fact that one has to pay close attention to growth properties of nonlinear quantities ψ, and on the other hand the assumption of weak convergence of $\{\psi(\nabla u_j)\}$ in $L^1(\Omega)$ in order to have representation of the limit in terms of the underlying $W^{1,p}$-parametrized measure is not automatically guaranteed. In the statement that follows $\mathcal{E}^p$ is essentially (not quite) the space of functions with growth of order at most p

$$|\varphi(A)| \leq C\,(1 + |A|^p)\,.$$

Theorems 8.14, 8.16 *Let $\nu = \{\nu_x\}_{x\in\Omega}$ be a family of probability measures supported on the space of matrices $\mathbf{M}^{m\times N}$. ν is a $W^{1,p}$-parametrized measure if and only if*

i) there exists $u \in W^{1,p}(\Omega)$ such that $\nabla u(x) = \int_{\mathbf{M}^{m\times N}} A\, d\nu_x(A)$, for a.e. $x \in \Omega$;

ii) $\int_{\mathbf{M}^{m\times N}} \varphi(A)\, d\nu_x(A) \geq \varphi(\nabla u(x))$ for every $\varphi \in \mathcal{E}^p$ quasiconvex and bounded from below and a.e. $x \in \Omega$;

iii) $\int_\Omega \int_{\mathbf{M}^{m\times N}} |A|^p\, d\nu_x(A)\, dx < \infty$.

For $p = \infty$ ii) and iii) should be changed to

ii') $\int_{\mathbf{M}^{m\times N}} \varphi(A)\, d\nu_x(A) \geq \varphi(\nabla u(x))$ for a.e. $x \in \Omega$ and any φ bounded from below and quasiconvex;

iii') $\operatorname{supp} \nu_x \subset K$ for a.e. $x \in \Omega$ where $K \subset \mathbf{M}^{m\times N}$ is some fixed bounded set.

It is worthwhile to mention some other crucial results used in proving this characterization in terms of Jensen's inequality. They are also important from the point of view of applications as well as interesting on their own. The first is the averaging procedure that has already been mentioned.

Theorem 8.1 *Let $\{u_j\}$ be a bounded sequence of functions in $W^{1,p}(\Omega)$ with affine boundary values given by u_Y. Let $\nu = \{\nu_x\}_{x\in\Omega}$ be the parametrized measure associated to $\{\nabla u_j\}$. There exists a sequence, $\{w_j\}$, bounded in $W^{1,p}(\Omega)$ with the same boundary values, such that the corresponding parametrized measure is $\overline{\nu}$, homogeneous, given by*

$$\langle \overline{\nu}, \varphi \rangle = \frac{1}{|\Omega|} \int_\Omega \int_{\mathbf{M}^{m\times N}} \varphi(A)\, d\nu_x(A)\, dx.$$

Another natural, important fact is that the quasiconvexification of a function is itself a quasiconvex function.

Lemma 8.10 *Let φ be an upper semicontinuous function defined on matrices with values in $\mathbf{R}^*$. The quasiconvexification of φ is a quasiconvex function.*

One important ingredient in proving weak lower semicontinuity for the case p finite is the following remarkable lemma.

Lemma 8.15 *Let $\{v_j\}$ be a bounded sequence in $W^{1,p}(\Omega)$. There always exists another sequence $\{u_j\}$ of Lipschitz functions ($u_j \in W^{1,\infty}(\Omega)$ for all j) such that $\{|\nabla u_j|^p\}$ is equiintegrable and the two sequences of gradients, $\{\nabla u_j\}$ and $\{\nabla v_j\}$, have the same underlying $W^{1,p}$-parametrized measure.*

Concerning relaxation we assumed in our arguments in the last section that

$$\psi^{\mathcal{L}}(u(x)) = \int_{\mathbf{R}^m} \psi(\lambda)\, d\nu_x(\lambda), \quad \nu_x \in \mathcal{A}_{u(x)}.$$

This is a question of attainment that is true for the gradient case under suitable hypotheses (Lemma 4.2). We will deal with relaxation in Chapter 4.

In the same spirit as the quasiconvexity condition (1–10), one can study the condition

$$Q^p\psi(Y) = \inf\left\{\frac{1}{|\Omega|}\int_\Omega \psi(Y + \nabla u(x))\, dx : u \in W_0^{1,p}(\Omega)\right\}$$

for finite p. If a function ψ defined on matrices is such that $Q^p\psi = \psi$ we say that it is $W^{1,p}$-quasiconvex. The reason why in our context, under growth conditions for ψ, we need not be concerned about this notion of convexity is established in the following statement.

Lemma 8.13 *Let ψ be a continuous function defined on matrices such that*

$$c \le \psi(A) \le C(1 + |A|^p), \quad C > 0, p \ge 1.$$

ψ is $W^{1,p}$-quasiconvex if and only if ψ is quasiconvex.

This whole program can be carried out for the operator div as well. This is the topic of Chapter 10. We restrict attention to the particular case of vector fields $u : \Omega \subset \mathbf{R}^N \to \mathbf{R}^{Nm}$,

$$\operatorname{div} u = \sum_{k=1}^N \frac{\partial u_k^i}{\partial x_k}, \quad i = 1, \dots, m,$$

and we take $m < N$. This case is particularly interesting from the point of view of some applications. It is also an example where real differential constraints give rise to no weaker restrictions on the convexity properties for integrands of weak lower semicontinuous functionals. Said differently, $\mathcal{L}$-convexity reduces in this case ($m < N$) to plain convexity.

Theorem 10.3 *Let $\nu = \{\nu_x\}_{x\in\Omega}$ be a family of probability measures depending measurably on $x \in \Omega \subset \mathbf{R}^N$ such that supp$(\nu_x) \subset K$ for a fixed compact $K \subset \mathbf{R}^m$. If $m < N$, there exists a sequence $u^j : \Omega \subset \mathbf{R}^N \to K$ such that $\{\operatorname{div}(u^j)\}$ is compact in $H^{-1}(\Omega)$ and its parametrized measure is ν.*

5. Some general facts about parametrized measures

This section is again a summary of a few basic, general, important facts to be utilized in proving results for all other chapters. Once again we refer to Chapter 6 for proofs and further remarks and discussions.

The basic existence theorem for parametrized measures that is sufficient for most of the situations of interest is

Theorem 6.2 *Let $\Omega \subset \mathbf{R}^N$ be a measurable set and let $z_j : \Omega \to \mathbf{R}^m$ be measurable functions such that*

$$\sup_j \int_\Omega g(|z_j|)\, dx < \infty,$$

where $g : [0,\infty) \to [0,\infty]$ is a continuous, non-decreasing function such that $\lim_{t\to\infty} g(t) = \infty$. There exists a subsequence, not relabeled, and a family of probability measures, $\nu = \{\nu_x\}_{x\in\Omega}$ (the associated parametrized measure), depending measurably on x, with the property that whenever the sequence $\{\psi(x, z_j(x))\}$ is weakly convergent in $L^1(\Omega)$ for any Carathéodory function $\psi(x,\lambda) : \Omega \times \mathbf{R}^m \to \mathbf{R}^*$, the weak limit is the (measurable) function

$$\overline{\psi}(x) = \int_{\mathbf{R}^m} \psi(x,\lambda)\, d\nu_x(\lambda).$$

By a Carathéodory function $\psi(x,\lambda) : \Omega \times \mathbf{R}^m \to \mathbf{R}^*$ we simply mean a function measurable in x and continuous in λ. Taking $g(t) = t^p$ or $g(t) = +\infty$ for $t > R$ we recover parametrized measures associated to bounded sequences in $L^p(\Omega)$ or $L^\infty(\Omega)$, respectively. Since Radon measures in $\mathbf{R}^m$ are identified by their action on continuous functions (or continuous functions vanishing at infinity), the parametrized measure associated to a sequence $\{z_j\}$ is determined by the weak limits of $\{\varphi(z_j)\}$ for all φ continuous vanishing at infinity. This is an important technical remark to bear in mind.

Chacon's biting lemma is a convenient tool when dealing with weak convergence in $L^1(\Omega)$. It enables us to remove the "bad" sets where concentrations occur, and have weak convergence on their complements.

Theorem 6.6 *(Chacon's biting lemma) Let $\{f_j\}$ be a uniformly bounded sequence in $L^1(\Omega)$,*

$$\sup_j \|f_j\|_{L^1(\Omega)} = C < \infty.$$

There exists a subsequence, not relabeled, a non-increasing sequence of measurable sets $\Omega_n \subset \Omega$, $|\Omega_n| \searrow 0$ and $f \in L^1(\Omega)$ such that

$$f_j \rightharpoonup f \text{ in } L^1(\Omega \setminus \Omega_n)$$

for all n.

The next result will also be important in proving weak lower semicontinuity facts if we have hypotheses in the form of Jensen's inequalities for parametrized measures. It explains, together with Lemma 8.15, why concentrations do not play a role in this type of variational principle: under growth assumptions we can always cut-off concentrations without increasing the energy.

Theorem 6.11 *If $\{z_j\}$ is a sequence of measurable functions with associated parametrized measure $\nu = \{\nu_x\}_{x\in\Omega}$, then*

$$\liminf_{j\to\infty} \int_E \psi(x, z_j(x))\, dx \geq \int_E \int_{\mathbf{R}^m} \psi(x,\lambda)\, d\nu_x(\lambda)\, dx,$$

for every Carathéodory function ψ, bounded from below, and every measurable subset $E \subset \Omega$.

Finally the next two propositions refer to how strong convergence is reflected on the parametrized measure.

Proposition 6.12 *Let $\{z_j\}$ be a sequence in $L^p(\Omega)$ such that $\{|z_j|^p\}$ is weakly convergent in $L^1(\Omega)$ for $p < \infty$ and let $\nu = \{\nu_x\}_{x\in\Omega}$ be the associated parametrized measure. $z_j \to z$ strongly in $L^p(\Omega)$ if and only if $\nu_x = \delta_{z(x)}$ for a.e. $x \in \Omega$.*

Proposition 6.13 *Let $z_j = (u_j, v_j) : \Omega \to \mathbf{R}^d \times \mathbf{R}^m$ be a bounded sequence in $L^p(\Omega)$ such that $\{u_j\}$ converges strongly to u in $L^p(\Omega)$. If $\nu = \{\nu_x\}_{x\in\Omega}$ is the parametrized measure associated to $\{z_j\}$ then $\nu_x = \delta_{u(x)} \otimes \mu_x$ a.e. $x \in \Omega$, where $\{\mu_x\}_{x\in\Omega}$ is the parametrized measure corresponding to $\{v_j\}$.*

We will apply this proposition to sequences $\{u_j, \nabla u_j\}$ for $\{u_j\}$ a weakly convergent sequence in $W^{1,p}(\Omega)$. In this situation we know that the functions themselves converge strongly to the weak limit by the Compactness Theorem of Sobolev spaces. If $u \in W^{1,p}(\Omega)$ is the weak limit and $\{\mu_x\}_{x\in\Omega}$ is the parametrized measure associated to the gradients $\{\nabla u_j\}$ then $\nu_x = \delta_{u(x)} \otimes \mu_x$ for a.e. $x \in \Omega$.

6. Appendix

Since Jensen's inequality will play a central role in these notes we include here a basic proof of the classical inequality for convex functions.

Theorem 1.2 *Let μ be a positive, Radon measure over a σ-algebra, $\mathcal{M}$, in a set Ω such that $\mu(\Omega) = 1$. Let f be a vector-valued function in $L^1(\mu)$ such that $f(x) \in K$ for μ-a.e. $x \in \Omega$ where $K \subset \mathbf{R}^m$ is a convex set. If φ is a convex function defined in K then*

$$\varphi\left(\int_\Omega f\, d\mu\right) \leq \int_\Omega \varphi(f)\, d\mu. \tag{1–12}$$

Proof. Since K is convex and $\mu(\Omega) = 1$, it is clear that

$$t = \int_\Omega f \, d\mu \in K.$$

The convexity of φ implies that we can find a $\beta \in \mathbf{R}^m$ such that

$$\varphi(s) \geq \varphi(t) + \beta \cdot (s - t), \quad s \in K.$$

In particular for μ-a.e. $x \in \Omega$

$$\varphi(f(x)) \geq \varphi(t) + \beta \cdot (f(x) - t),$$

and integrating over Ω, we get (1–12). (Keep in mind that $\mu(\Omega) = 1$.) ∎

We also remember in this section a few general facts about weak convergence in $L^p(\Omega)$.

Let $1 \leq p < \infty$. We say that $\{f_j\}$ converges weakly to f in $L^p(\Omega)$ and write $f_j \rightharpoonup f$ in $L^p(\Omega)$ if

$$\int_\Omega f_j(x) g(x) \, dx \to \int_\Omega f(x) g(x) \, dx,$$

for every $g \in L^q(\Omega)$, $1/p + 1/q = 1$. For $p = \infty$, $\{f_j\}$ converges weakly * to f in $L^\infty(\Omega)$ (in written form $f_j \overset{*}{\rightharpoonup} f$ in $L^\infty(\Omega)$) if

$$\int_\Omega f_j(x) g(x) \, dx \to \int_\Omega f(x) g(x) \, dx,$$

for every $g \in L^1(\Omega)$.

The criterion for weak compactness is contained in the following proposition. The case $p = 1$ is very special.

Proposition 1.3 *Let $1 < p \leq \infty$. The sequence $\{f_j\}$ is weakly relatively compact in $L^p(\Omega)$ (weakly * relatively compact if $p = \infty$) if and only if there exists a constant $K \geq 0$ such that $\|f_j\|_{L^p(\Omega)} \leq K$ uniformly for all j.*

Let $p = 1$. The sequence $\{f_j\}$ is relatively compact in $L^1(\Omega)$ if and only if:

i) there exists a constant $K \geq 0$ such that $\|f_j\|_{L^1(\Omega)} \leq K$ for all j;

ii) for every $\epsilon > 0$ there exists a $\delta = \delta(\epsilon) > 0$ such that for every measurable subset E with $|E| < \delta$, we have

$$\int_E |f_j(x)| \, dx < \epsilon,$$

uniformly for all j.

Condition ii) is the equiintegrability property or Dunford-Pettis criterion of weak compactness in $L^1(\Omega)$. It is a well-established fact that $L^1(\Omega)$ is very peculiar from the point of view of weak convergence and that peculiarity is precisely condition ii) above. Concentrations effects are connected to the failure of ii).

Proposition 1.4 *Let* $1 < p \leq \infty$. $f_j \rightharpoonup f$ *in* $L^p(\Omega)$ *(or weak * in* $L^\infty(\Omega)$*) if and only if*

i) $\|f_j\|_{L^p(\Omega)} \leq K$, $K > 0$;

ii) $\lim_{j\to\infty} \int_D (f_j(x) - f(x))\, dx = 0$, *for all cubes* $D \subset \Omega$.

Let $p = 1$. $f_j \rightharpoonup f$ *in* $L^1(\Omega)$ *if and only if*

i) $\|f_j\|_{L^1(\Omega)} \leq K$, $K > 0$;

ii) the equiintegrability property holds;

iii) $\lim_{j\to\infty} \int_D (f_j(x) - f(x))\, dx = 0$, *for all cubes* $D \subset \Omega$.

These notions of weak convergence in the $L^p(\Omega)$-spaces can be translated to the Sobolev spaces $W^{1,p}(\Omega)$. As usual we take

$$W^{1,p}(\Omega) = \{f \in L^p(\Omega) : \nabla f \in L^p(\Omega)\}.$$

Weak convergence in $W^{1,p}(\Omega)$ means weak convergence in $L^p(\Omega)$ for the functions and their gradients.

7. Bibliographical remarks

Many textbooks dealing with the calculus of variations are available. Some of them are less advanced ([7], [310]), broader in scope ([41], [70], [153], [157], [294]) or follow a more classical approach ([89], [158], [309]) while others are based on the direct method and on weak convergence techniques ([61], [98], [239], [313], [316]). An important complement on variational methods is convex analysis. We have included two basic such references [118], [274].

There are also many books on weak convergence. Most of them are graduate texts in functional analysis or partial differential equations. [6], [56], [115], [116], [132], [160], [284] are basic references. In the context of parametrized measures, where some of the examples in Section 1 can be found, [302] must be looked at. As an introduction to weak convergence, nonlinear functionals and parametrized measures the reader may try [93] as well.

The subject of parametrized measures is not new, especially for optimal control experts. There are a number of works on parametrized measures treated from a rather general and (for some of them) abstract point of view. These include [24], [27], [212], [213], [282], [314], [315] and the ones already cited.

Most of the ideas related to compensated compactness and the general framework described in Section 2 are scattered through the literature. Some of the fundamental sources for compensated compactness and the theory of homogenization are [247], [248], [249], [250], [303], [304], [306]. Applications of these methods to partial differential equations have been quite successful. In addition to the previous references, the reader is referred to [110], [111], [112], [113], [114], [265], [277], [286], [288]. These are included here as a sample since there exists a very large amount of work dealing with techniques in compensated compactness and homogenization. A nice account of the use of weak convergence in nonlinear partial differential equations is [132]. Parametrized measures have been used in evolution problems as well. Some examples in this direction are contained in [107], [149], [150], [178], [190], [290]. One of the main drawbacks of parametrized measures is that since they keep track of statistical properties alone, they are not well suited for problems in mathematical physics in which transport properties, multiscale phenomena, or more precise information concerning oscillations, play a fundamental role. To solve these difficulties H-measures have been introduced in [154], [155], [305]. Another point is that parametrized measures have not been designed to detect concentrations since they completely miss this effect on sequences. Several references concentrate on this issue [9], [145], [203], [204]. Additional references on compensated compactness and parametrized measures related to variational principles are [79], [80], [142], [255], [258].

There are topics beyond the scope of this book that are important from the point of view of variational principles, integral representations, functions of bounded variations, Γ-convergence, etc. A few basic references for these subjects are [18], [103], [161].

The proof of the classical Jensen's inequality for convex functions and probability measures included here is the one in the excellent book [283]. Other references on measure theory, functional analysis, partial differential equations, etc, where background material can be studied are [47], [90], [135], [164], [214], [268], in addition to the ones mentioned in the second paragraph of this section.

Most of the references on parametrized measures call them Young measures after the pioneering work of Young. We take the two terms as equivalent in this text.

Chapter 2
Some Variational Problems

1. Introduction

This chapter gathers a collection of problems for which the analysis does not involve any differential constraint, or if it does it is in a somewhat elementary way. It is a good way of practicing with the general ideas we will pursue for more complicated situations in subsequent chapters. For this reason we do not pretend to give the sharpest hypotheses under which theorems can be proved or improved, but rather focus on understanding the main techniques in each example. Some formal proofs are left to the reader as exercises. The same principle explains why in some of the problems we do not pursue the proof of all the steps and lemmas used when they are not relevant to our discussion. Three of the examples refer to variational principles or optimization. The last one does not. This has been included with the sole purpose of providing an illustration of how some analysis in terms of parametrized measures can also be helpful and provide some insight even though the problem is not directly related to variational principles but it is placed in a completely different context: large time behavior of complicated turbulent systems.

The first example in Section 2 corresponds exactly to the case of no differential constraint in the context of Chapter 1, Section 3: $\mathcal{L}$ is the set of bounded sequences in some $L^p(\Omega)$. The associated notion of convexity is the usual convexity so that Jensen's inequality does not place any real restriction on families of probability measures. A complete proof of a characterization theorem for the case $p = \infty$ is given. It is interesting to realize how differential constraints included in $\mathcal{L}$ make the analysis so much harder: the entire Chapter 8 (based on Chapter 6) is devoted to the characterization of parametrized measures under the curl constraint. In Section 3, a short discussion of simple optimal control problems is considered. Existence theorems and relaxation are described and examined in terms of parametrized measures associated to pairs of controls and states. We restrict ourselves to control problems governed by ordinary differential equations. Hypotheses are not completely explicit again because we want to emphasize the underlying analysis. Finally, an optimal design problem is described and a solution given in the form of parametrized measures. We are looking for the optimal design of a plate under the action of an external load. The optimality criterion is to minimize the compliance of the plate taken as the work done by the load. Minimizing sequences develop finer and finer oscillations but the weak limit of such sequences do not furnish the optimal design sought. Some generalized relaxed functional needs to be considered where families of probability measures compete. We will go back to this phenomenon of very fine spatial oscillations in the more involved applications of Chapter 5. Section 5 contains a short discussion of coherent structures and small-scale fluctuations in some turbulent systems.

2. Variational problems under no differential constraint

This section aims to be an introduction to variational problems in which there is no derivative or differential constraint involved. The prototype of such problems is

$$I(u) = \int_\Omega \varphi(x, u(x))\, dx, \tag{2–1}$$

where competing functions $u :. \Omega \subset \mathbf{R}^N \to \mathbf{R}^m$ are vector valued functions in some $L^p(\Omega)$ space. Ω is assumed to be a regular domain and $\varphi(x,\lambda) : \mathbf{R}^N \times \mathbf{R}^m \to \mathbf{R}^*$ is some known integrand assumed to be measurable in x and continuous in λ. We further consider the typical volume constraint

$$\int_\Omega u(x)\, dx = \alpha, \tag{2–2}$$

where $\alpha \in \mathbf{R}^m$ is given a priori. The most basic interpretation of this problem (for $m = 1$, the scalar case) consists in finding the best mass distribution (or the distribution of some other quantity), u, according to the optimization criterion of minimizing the integral (2–1) for a total amount of mass α given in (2–2). For the vector case $m > 1$ we are looking for the best distribution of several constituents for a fixed amount of each of them.

Suppose we want to find optimal solutions (minimizers) of our problem (2–1) and (2–2) through the direct method as described in Chapter 1. We need to understand the conditions on the density φ that ensure the weak lower semicontinuity property in $L^p(\Omega)$. Through coercivity assumptions on φ of the type

$$c\,(|\lambda|^p - 1) \le \varphi(x,\lambda), \quad p > 1, c > 0, \tag{2–3}$$

minimizing sequences will be uniformly bounded in $L^p(\Omega)$. Consider

$$\mathcal{L} = \{\{u_j\} \text{ bounded in } L^p(\Omega)\},$$

and define for any function $\psi : \mathbf{R}^m \to \mathbf{R}^*$,

$$\psi^{\mathcal{L}}(\lambda) = \inf\left\{\liminf_{j\to\infty} \frac{1}{|\Omega|}\int_\Omega \psi(u_j(x))\, dx : u_j \rightharpoonup \lambda \text{ in } L^p(\Omega)\right\}.$$

If $\lambda = t\lambda_1 + (1-t)\lambda_2$, after the examples in Chapter 1, it should be clear that we can find a bounded sequence in $L^\infty(\Omega)$, $\{u_j\}$, whose parametrized measure is precisely $\nu = t\delta_{\lambda_1} + (1-t)\delta_{\lambda_2}$, homogeneous. Indeed we can take

$$u_j(x) = \lambda_2 + \chi_t(jx \cdot n)\,(\lambda_1 - \lambda_2)$$

for any vector $n \in \mathbf{R}^N$, where χ_t is the characteristic function of $(0,t)$ in $(0,1)$ extended by periodicity. Therefore

$$\psi^{\mathcal{L}}(\lambda) \le t\psi(\lambda_1) + (1-t)\psi(\lambda_2).$$

Likewise, by induction, if

$$\lambda = \sum_i t_i\lambda_i, \quad t_i \ge 0, \sum_i t_i = 1,$$

there exists a bounded sequence in $L^\infty(\Omega)$, $\{u_j\}$, whose corresponding parametrized measure is

$$\nu = \sum_i t_i\delta_{\lambda_i},$$

and thus

$$\psi^{\mathcal{L}}(\lambda) \le \sum_i t_i\psi(\lambda_i).$$

By taking the infimum in the right-hand side we arrive at

$$\psi^{\mathcal{L}}(\lambda) \le \psi^{**}(\lambda),$$

where ψ^{**} is the usual convexification of ψ. The reverse inequality is also true. By Jensen's inequality

$$\begin{aligned}\frac{1}{|\Omega|}\int_\Omega \psi(u_j(x))\,dx &\ge \frac{1}{|\Omega|}\int_\Omega \psi^{**}(u_j(x))\,dx \\ &\ge \psi^{**}\left(\frac{1}{|\Omega|}\int_\Omega u_j(x)\,dx\right).\end{aligned}$$

The limit of the right-hand side is $\psi^{**}(\lambda)$ for any sequence $\{u_j\}$ converging weakly to λ. In this case $\psi^{\mathcal{L}} = \psi^{**}$ and all the required assumptions in the framework described in Chapter 1, Section 3 are trivially true since Jensen's inequality with respect to all convex functions does not place any restriction on families of probability measures.

Theorem 2.1 *Let $\nu = \{\nu_x\}_{x\in\Omega}$ be a family of probability measures depending measurably on $x \in \Omega$ such that supp$(\nu_x) \subset K \subset \mathbf{R}^m$, a.e. $x \in \Omega$. There exists a sequence, $\{u_j\}$, bounded in $L^p(\Omega)$ and taking values in K, whose parametrized measure is ν if and only if*

$$\int_\Omega \int_K |\lambda|^p \, d\nu_x(\lambda)\,dx < \infty, \tag{2–4}$$

for $p < \infty$, and if and only if K is bounded for $p = \infty$.

Proof. We provide the details for the case $p = \infty$ in which we assume that K is bounded. The extension to the finite case $p < \infty$ requires approximation and truncation techniques that are not relevant to the main stream of our discussion. They will be treated in Chapter 7.

If there exists a sequence, $\{u_j\}$, bounded in $L^p(\Omega)$, the associated parametrized measure satisfies (2–4), by Theorem 6.11. The restriction on the support of each ν_x is also elementary if the image for each u_j lies inside K. The converse is the interesting part of the theorem.

We place ourselves in the Banach space $\mathcal{M}(\overline{\Omega} \times K)$ of Radon measures on $\overline{\Omega} \times K$ under the total variation norm, where we assume Ω to be bounded and K compact, so that the product $\overline{\Omega} \times K$ is compact. In this case this Banach space is the dual of $\mathcal{C}(\overline{\Omega} \times K)$ endowed with the sup norm which is separable. Convergence in $\mathcal{M}(\overline{\Omega} \times K)$ with respect to the weak * topology can be characterized by sequences. Consider

$$\mathcal{A} = \Big\{ \mu_u \in \mathcal{M}(\overline{\Omega} \times K) : \langle \mu_u, \psi \rangle = \int_\Omega \psi(x, u(x))\, dx,$$
$$\psi \in \mathcal{C}(\overline{\Omega} \times K), u : \Omega \to K \Big\},$$

and let $\mu \in \mathcal{M}(\overline{\Omega} \times K)$ be defined by

$$\langle \mu, \psi \rangle = \frac{1}{|\Omega|} \int_\Omega \int_K \psi(x, \lambda)\, d\nu_x(\lambda)\, dx,$$

where the family $\nu = \{\nu_x\}_{x \in \Omega}$ is given verifying $\operatorname{supp}(\nu_x) \subset K$ a.e. $x \in \Omega$. For the case $p = \infty$, (2–4) drops out and is replaced by the boundedness restriction on K.

Step 1. $\overline{\mathcal{A}}$ is a convex set (closure is meant in the weak * topology). Let $t \in (0,1)$ and let u_i, $i = 1, 2$ be measurable functions taking values in K. We would like to show that the measure $t\mu_{u_1} + (1-t)\mu_{u_2}$ belongs to $\overline{\mathcal{A}}$. Let once again χ_t stand for the characteristic function of $(0,t)$ in $(0,1)$ extended by periodicity to all of $\mathbf{R}$, and write $\chi_k(x) = \chi_t(kx \cdot n)$ for any non-zero vector $n \in \mathbf{R}^N$. As we have argued in Chapter 1, $\chi_k \overset{*}{\rightharpoonup} t$ in $L^\infty(\Omega)$. Consider

$$u^{(k)}(x) = \chi_k(x) u_1(x) + (1 - \chi_k(x))\, u_2(x).$$

We claim that $\mu_{u^{(k)}} \overset{*}{\rightharpoonup} t\mu_{u_1} + (1-t)\mu_{u_2}$ in $\mathcal{M}(\overline{\Omega} \times K)$. In fact for any continuous ψ,

$$\begin{aligned}\lim_{k\to\infty} \int \psi(x, u^{(k)}(x))\, dx &= \lim_{k\to\infty} \int_\Omega \psi\left(x, \chi_k(x)u_1(x) + (1 - \chi_k(x))\, u_2(x)\right) dx \\ &= \lim_{k\to\infty} \int_\Omega [\chi_k(x)\psi(x, u_1(x)) \\ &\qquad + (1 - \chi_k(x))\, \psi(x, u_2(x))]\, dx \\ &= \langle t\mu_{u_1} + (1-t)\mu_{u_2}, \psi \rangle.\end{aligned}$$

Step 2. $\mu \in \overline{\mathcal{A}}$. For this we use the Hahn-Banach theorem. Suppose T is a linear functional represented by a particular function $\psi \in \mathcal{C}(\overline{\Omega} \times K)$ such that $\langle \mu_u, \psi \rangle \geq 0$ for all measurable $u : \Omega \to K$:

$$\int_\Omega \psi(x, u(x))\, dx \geq 0. \tag{2–5}$$

Take $u(x) = \min_\lambda \psi(x, \lambda)$. Since $\psi(x, \lambda) \geq \psi(x, u(x))$ and μ is non-negative,

$$\begin{aligned}\int_{\Omega \times K} \psi(x, \lambda)\, d\mu(x, \lambda) &\geq \int_{\Omega \times K} \psi(x, u(x))\, d\mu(x, \lambda) \\ &= \int_\Omega \psi(x, u(x))\, dx \\ &\geq 0,\end{aligned}$$

by (2–5).

Step 3. Conclusion. Because of the remark made earlier about how sequences characterize weak * convergence in $\mathcal{M}(\overline{\Omega} \times K)$, by Step 2 we can find a sequence of measurable functions, $\{u_j\}$, such that

$$\lim_{j \to \infty} \int_\Omega \psi(x, u_j(x))\, dx = \int_\Omega \int_K \psi(x, \lambda)\, d\nu_x(\lambda)\, dx,$$

for all continuous ψ. In particular, if $\varphi : K \to \mathbf{R}$ is continuous the sequence (or some suitable subsequence) $\{\varphi(u_j)\}$ will converge, weakly * in the sense of measures, to

$$\overline{\varphi}(x) = \int_K \varphi(\lambda)\, d\nu_x(\lambda).$$

By uniqueness of the limit $\varphi(u_j) \overset{*}{\rightharpoonup} \overline{\varphi}$ in $L^\infty(\Omega)$ so that $\nu = \{\nu_x\}_{x \in \Omega}$ is the parametrized measure associated to $\{u_j\}$. ■

Based on this characterization, we can proceed to analyze weak lower semicontinuity and relaxation for I in (2–1) along the lines developed in Chapter 1, Section 3. The reader is invited to provide the details. Notice that the constraint on the total mass (2–2) is preserved under weak convergence.

3. Optimal control problems

Optimal control is a part of the theory of optimization more general than the calculus of variations. We would like to study as an example one of the most basic optimal control problems governed by ordinary differential equations in order to show how parametrized measures may serve to analyze this type of problems as well. As a matter of fact, our general framework in Chapter 1 is also useful in this context.

Our (payoff) functional I is of the form

$$I(u,y) = \int_J \varphi(t, u(t), y(t))\, dt$$

where J is some interval (finite of infinite) of $\mathbf{R}$, $u : J \to \mathbf{R}^m$ is the control variable (the free variable) and $y : J \to \mathbf{R}^d$ is the state of the particular system under consideration coupled to the control through the equation of state

$$y'(t) = A(t, u(t), y(t)),$$

where $A : J \times \mathbf{R}^m \times \mathbf{R}^d \to \mathbf{R}^d$ is such that existence of solutions to the equation of state are guaranteed for the class of controls we want to consider. There might be some other constraints in the problem like $u(t) \in K$ for some fixed subset $K \subset \mathbf{R}^m$ or restrictions on initial conditions for the equation of state. For definiteness we neglect these other conditions, or assume them to be preserved by weak convergence otherwise. This last hypothesis might not be true, though, in some circumstances of interest and may require some further analysis.

Assume that $\varphi : J \times \mathbf{R}^m \times \mathbf{R}^d \to \mathbf{R}$ is continuous in all its arguments and we have the coerciveness hypothesis

$$c\left(|u|^p + |y|^p - 1\right) \le \varphi(t,u,y), \quad p > 1, c > 0,$$

for all $(t,u,y) \in J \times \mathbf{R}^m \times \mathbf{R}^d$. Minimizing sequences will be bounded in $L^p(J)$ under this assumption. We consider

$$\mathcal{L} = \left\{\{(u_j, y_j)\} \text{ bounded in } L^p(J) : y_j' = A(t, u_j, y_j)\right\}.$$

Assume further that

$$|A(t,u,y)| \le C\left(|u|^q + |y|^q + 1\right), \quad 1 \le q < p.$$

In this case, if $\{(u_j, y_j)\} \in \mathcal{L}$ and $\mu = \{\mu_t\}_{t \in J}$ is the associated underlying parametrized measure, then $\left\{|A(t, u_j, y_j)|^{p/q}\right\}$ is bounded, and since $q < p$, $\{A(t, u_j, y_j)\}$ (or some appropriate subsequence) converges weakly in $L^1(J)$. Consequently,

$$y_j' = A(t, u_j, y_j) \rightharpoonup \int_{\mathbf{R}^m \times \mathbf{R}^d} A(t, \lambda_1, \lambda_2)\, d\mu_t(\lambda_1, \lambda_2) = \overline{A}(t),$$

where $\overline{A} \in L^1(J)$. If we define

$$y(t) = \int_0^t \overline{A}(s)\, ds,$$

modulo a constant, $y_j \to y$ strong in $L^\infty(J)$. Recalling the comments about how strong convergence is reflected on the parametrized measure, Proposition 6.13, we conclude that if $\nu = \{\nu_t\}_{t\in J}$ is the parametrized measure associated to the sequence of controls $\{u_j\}$ then

$$\mu_t = \nu_t \otimes \delta_{y(t)}, \quad \text{a.e. } t \in J, \tag{2–6}$$

and

$$\overline{A}(t) = \int_{\mathbf{R}^m} A(t, \lambda, y(t))\, d\nu_t(\lambda).$$

Sequences in $\mathcal{L}$ correspond to strongly convergent sequences in $L^\infty(J)$ for the state variable and weakly convergent sequences in $L^p(J)$ for the control variable. This in particular implies that the appropriate notion of convexity (associated to $\mathcal{L}$) for weak lower semicontinuity is usual convexity of φ with respect to the control variable: if $u_j \rightharpoonup u$ in $L^p(\Omega)$ and $y_j \to y$ in $L^\infty(\Omega)$ then under convexity of φ with respect to u and boundedness from below, by Theorem 6.11, (2–6) and Jensen's inequality, we have

$$\begin{aligned}\lim_{j\to\infty} \int_J \varphi(t, u_j(t), y_j(t))\, dt &\geq \int_J \int_{\mathbf{R}^m \times \mathbf{R}^d} \varphi(t, \lambda_1, \lambda_2)\, d\mu_t(\lambda_1, \lambda_2)\, dt \\ &= \int_J \int_{\mathbf{R}^m} \varphi(t, \lambda, y(t))\, d\nu_t(\lambda)\, dt \\ &\geq \int_J \varphi\left(t, \int_{\mathbf{R}^m} \lambda\, d\nu_t(\lambda), y(t)\right)\, dt \\ &= \int_J \varphi(t, u(t), y(t))\, dt,\end{aligned}$$

if a suitable subsequence has been chosen. This time, however, this condition does not ensure by itself the success of the direct method to achieve minimizers for our problem. Indeed, $\mathcal{L}$ is not weakly closed. To see this, suppose $\{(u_j, y_j)\}$ is a sequence in $\mathcal{L}$ so that $y_j \to y$ and $u_j \rightharpoonup u$. The crucial question is whether u and y are coupled by the equation of state. We know that if ν is the parametrized measure associated to the sequence of controls then

$$y'(t) = \int_{\mathbf{R}^m} A(t, \lambda, y(t))\, d\nu_t(\lambda),$$

so that for weak closedness we must require

$$\int_{\mathbf{R}^m} A(t, \lambda, y(t))\, d\nu_t(\lambda) = A(t, u(t), y(t)), \quad u(t) = \int_{\mathbf{R}^m} \lambda\, d\nu_t(\lambda).$$

This condition does not hold for all choices of ν unless A is linear in u. In the spirit of the discussion of Chapter 1, the reader can rigorously prove the following existence theorem for the optimal control problem.

Theorem 2.2 *Assume that the following hypotheses hold:*

i) $A(t,u,y) = A_1(t,y)u + A_2(t,y)$ *where* $A_1 : J\times\mathbf{R}^d \to \mathbf{M}^{m\times d}$, $A_2 : J\times\mathbf{R}^d \to \mathbf{R}^d$ *and*

$$|A_1(t,y)| \le C, \quad |A_2(t,y)| \le C\,(1+|y|)\,;$$

ii) φ *is continuous, convex in* u *and*

$$c\,(|u|^p + |y|^p - 1) \le \varphi(t,u,y), \quad p>1.$$

The associated optimal control problem admits a solution.

If A is not linear in u, even though φ may be convex on the control variable, the analysis might proceed seeking a relaxed or generalized functional defined on parametrized measures associated to sequences of controls

$$\overline{I}(\nu) = \int_J \int_{\mathbf{R}^m} \varphi(t,\lambda,y(t))\,d\nu_t(\lambda)\,dt,$$

where

$$y'(t) = \int_{\mathbf{R}^m} A(t,\lambda,y(t))\,d\nu_t(\lambda),$$

and $\nu = \{\nu_t\}_{t\in J}$ satisfies

$$\int_{\mathbf{R}^m} |\lambda|^p \; d\nu_t(\lambda) < \infty.$$

Again there might be more restrictions on ν coming from the additional initial constraints. There are, however, some technical difficulties to be overcome with this generalized formulation related to the differential equation for y which is written this time in terms of a family of probability measures.

4. An optimal design problem

We describe in this section some analysis of an optimal design problem for a plate of variable thickness under the model of Kirchhoff for pure bending of symmetric plates. We try to find the optimal structure with respect to the overall rigidity of the plate under the action of an external load. The model we consider is a somewhat simplified version where the thickness of the plate depends on just one variable and the tensors involved in the analysis depend upon the design of the plate through the half-thickness h. Let Ω be a regular, smooth domain in $\mathbf{R}^2$ representing the midplane of the plate with respect to which the plate is symmetric. The deflection or vertical displacement w in the model under consideration obeys the fourth order, elliptic equation

$$\frac{\partial^2}{\partial x_\alpha \partial x_\beta}\left(M_{\alpha\beta\gamma\delta}\frac{\partial^2 w}{\partial x_\gamma \partial x_\delta}\right) = F, \tag{2–7}$$

where $F \in L^2(\Omega)$ is the vertical load on the plate. The summation convention is used throughout this section. This equation must be satisfied in Ω. The design of the plate is hidden in the tensor $M_{\alpha\beta\gamma\delta}$ through the dependence

$$M_{\alpha\beta\gamma\delta} = \frac{2}{3} h^3(x_1, x_2) B_{\alpha\beta\gamma\delta},$$

where h is the thickness and $B_{\alpha\beta\gamma\delta}$ is a constant tensor that depends on material constants alone. In order to use Lemma 2.3 below, we have to restrict ourselves to the case where the thickness h is in fact a function of x_1 alone (though we will still write $h(x)$, $x \in \Omega$), and x_1 belongs to the interval

$$(a, b) = \{x_1 \in \mathbf{R} : \text{ there exists some } x_2 \in \mathbf{R} \text{ with } (x_1, x_2) \in \Omega\}.$$

We further restrict the class of materials by imposing a orthotropic condition: the nonzero components of $B_{\alpha\beta\gamma\delta}$ are

$$B_{1111} = B_{2222} = \frac{E}{1 - r^2},$$
$$B_{1122} = B_{2211} = \frac{Er}{1 - r^2},$$
$$B_{1212} = B_{1221} = B_{2112} = B_{2121} = \frac{E}{2(1 + r)},$$

where E and r stand for the Young's modulus and the Poisson ratio, respectively. Equation (2–7) is completed with the boundary conditions

$$w = \frac{\partial w}{\partial n} = 0, \quad \text{on } \partial\Omega, \tag{2–8}$$

reflecting the hypothesis that the plate is clamped. The boundary value problem (2–7) together with the boundary conditions (2–8) is variational, so that the solution is indeed the minimizer of the functional

$$\mathcal{E}(w) = \frac{1}{2} \int_\Omega M_{\alpha\beta\gamma\delta} \frac{\partial^2 w}{\partial x_\alpha \partial x_\beta} \frac{\partial^2 w}{\partial x_\gamma \partial x_\delta} \, dx - \int_\Omega F w \, dx,$$

(again the summation convention is assumed) over $H_0^2(\Omega)$, the subspace of $H^2(\Omega)$ satisfying (2–8). This can be easily checked. $H^2(\Omega)$ is the Hilbert space of $L^2(\Omega)$-functions having first and second weak derivatives in $L^2(\Omega)$.

The compliance of the plate is defined to be the work done by the load F and is regarded as a function of the half-thickness h,

$$L(h) = \int_\Omega F w \, dx. \tag{2–9}$$

It yields a measure of the rigidity or flexibility of the plate under the action of F. The design or optimization object is to minimize $L(h)$ among all the admissible plates with prescribed volume. The technical assumptions on the half-thicknesess h that may compete in (2–9) are the following

$$\mathcal{H} = \left\{ h \in L^\infty(\Omega) : h_{min} \leq h(x) \leq h_{max}, \int_\Omega h(x)\, dx = V_0 \right\},$$

where h_{min}, h_{max} and V_0 are prescribed a priori in a consistent way

$$0 < h_{min} |\Omega| < V_0 < h_{max} |\Omega| .$$

The basic feature of this optimization problem is the lack of minimizers. Minimizing sequences oscillate abruptly seeking the minimum value of the compliance available. In such cases a relaxation of the problem should be performed. What this amounts to is to provide some precise description, as simple as possible, of minimizing sequences. There might be many different types of minimizing sequences that realize the infimum of the compliance, some of them extremely complicated. To determine a relaxation is to search for a way to describe minimizing sequences with as few variables as possible. This description should be valid for all choices of the different parameters of the problem.

The basic tool to describe relaxation in this context is the following well-known lemma. It also explains why certain expressions (the cubic-average and harmonic cubic-average) arise in these relaxations. In order to state this result, we need some notation. A fourth order tensor $M(x)$ is said to be orthotropic if the non-vanishing coefficients are M_{1111}, M_{2222} and

$$M_{1122} = M_{2211}, \quad M_{1212} = M_{1221} = M_{2112} = M_{2121}.$$

M is bounded by the constants (d, D) if for every symmetric tensor $t = t_{\alpha\beta}$ we have for every $x \in \Omega$

$$d\,|t|^2 \leq M_{\alpha\beta\gamma\delta} t_{\alpha\beta} t_{\gamma\delta},$$
$$|M_{\alpha\beta\gamma\delta} t_{\alpha\beta}| \leq D\,|t| \quad \text{for every } \gamma, \delta.$$

Lemma 2.3 *Let $\{M^k\}$ be a sequence of orthotropic tensors bounded uniformly by (d, D). Let us assume that*

$$\left(M^k_{1111}\right)^{-1} \overset{*}{\rightharpoonup} \left(M^\infty_{1111}\right)^{-1},$$
$$\left(M^k_{1122}\right)\left(M^k_{1111}\right)^{-1} \overset{*}{\rightharpoonup} \left(M^\infty_{1122}\right)\left(M^\infty_{1111}\right)^{-1},$$
$$\left(M^k_{2222}\right) - \left(M^k_{1122}\right)^2\left(M^k_{1111}\right)^{-1} \overset{*}{\rightharpoonup} \left(M^\infty_{2222}\right) - \left(M^\infty_{1122}\right)^2\left(M^\infty_{1111}\right)^{-1},$$
$$M^k_{1212} \overset{*}{\rightharpoonup} M^\infty_{1212}.$$

If w^k, $1 \leq k \leq \infty$, is the solution of (2–7) and (2–8) corresponding to M^k, then $w^k \rightharpoonup w^\infty$ in $H^2_0(\Omega)$.

We examine relaxation directly in terms of parametrized measures and find easily a generalized minimizer. Once we achieve the existence of minimizers it is interesting to look for other minimizers, having in mind to simplify the understanding of minimizing sequences that generate such minimizers. In this sense, the motivation is to use as few design variables as possible to describe generalized minimizers. It should be noted that this process can be accomplished with this particular problem because the generalized compliance functional depends only upon certain moments of the parametrized measures associated to minimizing sequences.

Let $\overline{\mathcal{H}}$ be the set of parametrized measures associated to sequences h_k of half-thicknesses. In view of Theorem 2.1, the only restriction we have on such families is the support and the volume integral

$$\overline{\mathcal{H}} = \Big\{\mu = \{\mu_x\}_{x\in\Omega} : \operatorname{supp}\mu_x \subset Q = [h_{min}, h_{max}] \text{ a.e. } x \in \Omega, \\ \int_\Omega \int_Q \lambda \, d\mu_x(\lambda)\, dx = V_0\Big\}.$$

Notice that for any such μ we can find, according to Theorem 2.1, a sequence $\{h_k\}$ taking values in Q and whose associated parametrized measure is precisely μ. It might not be true, however, that

$$\int_\Omega h_k(x)\, dx = V_0, \quad \text{for all } k.$$

What we do know is that

$$\int_\Omega h_k(x)\, dx \to V_0.$$

To solve this problem is a pure technicality and involves changing each h_k in a small set without changing the parametrized measure. The reader is invited to provide the details. See Lemma 6.3 in Chapter 6 (this lemma has not been included in Chapter 1).

In order to define a compliance in $\overline{\mathcal{H}}$, let us further examine Lemma 2.3. The different weak limits we should care about in our case in order to apply the lemma are

$$\left(M^k_{1111}\right)^{-1} = \left(\frac{2}{3}h_k^3\frac{E}{1-r^2}\right)^{-1},$$

$$\left(M^k_{1122}\right)\left(M^k_{1111}\right)^{-1} = r,$$

$$\left(M^k_{2222}\right) - \left(M^k_{1122}\right)^2\left(M^k_{1111}\right)^{-1} = \frac{2}{3}h_k^3 E,$$

$$\left(M^k_{1212}\right) = \frac{1}{3}h_k^3\frac{E}{1+r}.$$

These weak limits can be represented through the moments of order 3 and -3 of the parametrized measure μ corresponding to the sequence $\{h_k\}$. Hence, if we let

$$\begin{aligned} m(x) &= \int_Q \lambda^3 \, d\mu_x(\lambda). \\ c^{-1}(x) &= \int_Q \lambda^{-3} \, d\mu_x(\lambda), \end{aligned} \tag{2–10}$$

and define

$$\begin{aligned} \left(M^\infty_{1111}\right)^{-1} &= \left(\frac{2}{3}c(x)\frac{E}{1-r^2}\right)^{-1}, \\ \left(M^\infty_{1122}\right)\left(M^\infty_{1111}\right)^{-1} &= r, \\ \left(M^\infty_{2222}\right) - \left(M^\infty_{1122}\right)^2\left(M^\infty_{1111}\right)^{-1} &= \frac{2}{3}m(x)E, \\ \left(M^\infty_{1212}\right) &= \frac{1}{3}m(x)\frac{E}{1+r}, \end{aligned} \tag{2–11}$$

by Lemma 2.3 (the other hypotheses in the lemma are easily verified in this situation), the displacements w_k corresponding to the tensors M^k associated in turn to h_k which generate μ, will converge weakly to the solution of the same problem with the tensor M^∞. Thus we must define the compliance $\overline{L}$ for elements in $\overline{\mathcal{H}}$ to be

$$\overline{L}(\mu) = \int_\Omega F\,w\,dx,$$

where w is the solution of (2–7), (2–8), with the tensor M^∞ depending on μ through (2–10) and (2–11).

Theorem 2.4

$$\inf_{\mathcal{H}} L = \min_{\overline{\mathcal{H}}} \overline{L}.$$

Proof. At this point the proof of the theorem has almost been indicated. First, notice that for $h \in \mathcal{H}$,

$$\mu = \delta_{h(x)} \in \overline{\mathcal{H}},$$

and, moreover, $L(h) = \overline{L}(\mu)$, so that

$$\inf_{\overline{\mathcal{H}}} \overline{L} \leq \inf_{\mathcal{H}} L.$$

On the other hand, given any $\mu \in \overline{\mathcal{H}}$ we can find a sequence $\{h_k\} \subset \mathcal{H}$ whose parametrized measure is μ, as indicated. Again by Lemma 2.3 we conclude the weak convergence of the solutions to (2–7), (2–8) as before, and thus

$$\overline{L}(\mu) = \lim_{k\to\infty} L(h_k).$$

The arbitrariness of μ yields the equality of the two infima.

To show existence of minimizers for $(\overline{\mathcal{H}}, \overline{L})$ is now an easy task. Take any minimizing sequence for L in $\mathcal{H}$. The parametrized measure generated by such sequence μ is admissible since it belongs to $\overline{\mathcal{H}}$ and by definition of $\overline{L}$ we have as before

$$\overline{L}(\mu) = \lim_{k\to\infty} L(h_k),$$

so that μ is truly a minimizer. ∎

A crucial observation is that $\overline{L}$ depends only upon the moments of order 3 and -3 of μ, in such a way that if μ^1 and μ^2 have in common these two moments then $\overline{L}(\mu^1) = \overline{L}(\mu^2)$. This brings us to the question of finding the easiest $\mu \in \overline{\mathcal{H}}$ that has the same moments of order 3 and -3 as a given minimizer μ whose existence is guaranteed in Theorem 2.4. Let us set

$$m(x) = \int_Q \lambda^3 \, d\mu_x(\lambda),$$

$$c^{-1}(x) = \int_Q \lambda^{-3} \, d\mu_x(\lambda),$$

where μ is a minimizer. Given Q, V_0, m and c, the problem reduces to seeking a family of probability measures as simple as possible whose support is contained in Q, whose integral volume is V_0 and whose moments of order 3 and -3 are m and c^{-1}, respectively. Any family verifying these conditions is a minimizer for $\overline{L}$ and therefore any generating sequence of such a parametrized measure will be a minimizing sequence for our original optimization problem.

Although it is beyond the purpose of this book, one can actually find minimizers for $\overline{L}$ of the form

$$\nu_x = \begin{cases} \theta(x)\delta_{h_{max}} + (1-\theta(x))\delta_{h(x)}, & x \in (a, \lambda), \\ \theta(x)\delta_{h_{min}} + (1-\theta(x))\delta_{h(x)}, & x \in (\lambda, b), \end{cases}$$

for some $\theta(x) \in [0, 1]$, $h \in \mathcal{H}$, and $\lambda \in [a, b]$. This generalized minimizer is the one that requires a minimal number of design variables.

5. Turbulent fluids

One of the most striking features of many turbulent fluid systems is the appearance of large-scale organized states, or *coherent structures*, in the midst of small-scale fluctuations. Such phenomena occurs, for example, in high Reynolds number two-dimensional hydrodynamics, and in slightly dissipative magnetofluids in two and three dimensions. The parametrized measure has proven to be a useful device in the modeling and analysis of coherent structures inherent in the long-evolved state of such systems. Roughly speaking, the parametrized measure represents a long-time weak limit of the relevant turbulent fluctuating fields, and the parametrized mean associated with this measure defines a macroscopic organized state. Here, we illustrate these methods, focusing on two-dimensional hydrodynamics.

The dynamics of an ideal, incompressible two-dimensional fluid is governed by the Euler equations:

$$\frac{\partial \omega}{\partial t} + u \cdot \nabla \omega = 0, \quad \omega(0, x) = \omega_0(x). \tag{2–12}$$

Here $u = (u_1, u_2)$ is the fluid velocity and

$$\omega = \nabla \times u = \frac{\partial u_2}{\partial x_1} - \frac{\partial u_1}{\partial x_2}$$

is the scalar vorticity field. The equations are assumed to hold in a bounded, simply connected spatial domain $\Omega \subset \mathbf{R}^2$ with smooth boundary $\partial\Omega$. The velocity field is divergence free, $\nabla \cdot u = 0$, and tangential $u \cdot n = 0$ on $\partial\Omega$. Consequently, there exists a stream function $\psi(x)$ such that

$$u = \left(\frac{\partial \psi}{\partial x_2}, -\frac{\partial \psi}{\partial x_1} \right),$$

with $\psi = 0$ on $\partial\Omega$. The stream function and vorticity are, therefore, related through the elliptic boundary value problem

$$-\Delta \psi = \omega, \text{ in } \Omega, \quad \psi|_{\partial\Omega} = 0, \tag{2–13}$$

and thus the vorticity transport equation (2–12) can be expressed entirely in terms of ω alone

$$\frac{\partial \omega}{\partial t} + \partial(\omega, G\omega) = 0, \tag{2–14}$$

where we have written $\psi = G\omega$ with G the Green's operator corresponding to the Dirichlet problem (2–13) and $\partial(f, g) = \det(\nabla f, \nabla g)$. The nonlinear scalar evolution equation (2–14) is known to be well-posed for bounded measurable vorticity functions. More precisely, if the initial vorticity satisfies $\omega_0 \in L^\infty(\Omega)$ then $\omega \in L^\infty((0,\infty) \times \Omega)$ and the $L^\infty(\Omega)$-norm of $\omega(t, \cdot)$ is preserved for all $t > 0$. This bound on vorticity provides adequate smoothness on the velocity field to guarantee existence and uniqueness of particle paths

$$\frac{dx}{dt} = u(t, x), \quad x(0) = x_0 \in \Omega,$$

from which it follows that there is a unique weak solution $\omega(t, x)$ of equation (2–14) for any initial vorticity $\omega_0 \in L^\infty(\Omega)$. Even for smooth initial vorticity fields, however, the regularity of weak solutions to (2–14) quickly degenerates as time proceeds, owing to the rapid growth of the vorticity gradient. This growth results from the increasingly intricate spatial arrangements realized by the vorticity field as it is advected by the flow. This turbulent behavior is well-documented by numerous direct numerical simulations of high Reynolds number two-dimensional flows. Because of its highly complicated microscopic behavior, the vorticity field $\omega(t,x)$ itself, therefore, does not provide a useful description of the long-time behavior of the fluid. For this reason, it is desirable to shift to a macroscopic description of the vorticity distribution that only partially encodes the rapidly increasing information content of the microscopic vorticity field. Such a description is afforded by the parametrized measure $\nu = \{\nu_x\}_{x\in\Omega}$ associated to the sequence of functions $\{\omega(t_j,\cdot)\}$ when $t_j \to \infty$ for a weak solution of the Euler equation, ω. Indeed, if $\|\omega_0\|_{L^\infty(\Omega)} = r$, then for all $t > 0$, $\|\omega(t,\cdot)\|_{L^\infty(\Omega)} \leq r$. Therefore, we can find sequences $t_j \to \infty$ such that $\{\omega(t_j,\cdot)\}$ generates a parametrized measure $\nu = \{\nu_x\}_{x\in\Omega}$ with the support contained in the interval $[-r,r]$. This measure captures the limiting statistics of the sequence $\{\omega(t_j,\cdot)\}$ in an infinitesimal neighborhood of each point in the spatial domain Ω.

There may be many such parametrized measures depending on the particular sequence t_j that is chosen. We wish, therefore, to select from the set of possible parametrized measure weak limits the one that is in some sense most likely to be realized. The first difficulty that is confronted in this program is that of determining an appropriate class of admissible parametrized measures. We must recognize that it is seemingly impossible to characterize completely the set of such families of measures that can be generated by sequences of vorticity functions corresponding to a solution of the Euler equations. This is due to the highly complex behavior exhibited by the vorticity field as it evolves, as alluded to above. Indeed, for all practical purposes, the only useful information that remains after a certain period of time is that the energy and entropy of the system are invariant under the dynamics. These quantities may be expressed as functionals on the vorticity field. The requisite formulas are, respectively,

$$E(\omega) = \frac{1}{2}\int_\Omega \omega\psi\,dx, \qquad F_f(\omega) = \int_\Omega f(\omega)\,dx,$$

where f can be any continuous function in $[-r,r]$. Notice that there is an infinite family of conserved entropy integrals. It is generally accepted that these are the only invariants of the dynamics, aside from those that may arise from special domain geometries. We shall assume that they exhaust the list of invariant functionals.

The conservation of energy and entropy by the dynamics translates into corresponding constraints on the possible parametrized measure weak limits. Indeed, if E^0 and F_f^0 are the values of energy and entropy fixed by the initial vorticity ω_0, then for $\omega_j = \omega(t_j,\cdot)$

$$\hat{E}(\nu) = \lim_{j\to\infty} E(\omega_j) = \frac{1}{2}\int_\Omega \hat{\omega}\hat{\psi}\,dx = E^0,$$

and

$$\hat{F}_f(\nu) = \lim_{j\to\infty} F_f(\omega_j) = \int_\Omega \int_{[-r,r]} f(y)\, d\nu_x(y)\, dx = F_f^0$$

where $\hat{\omega}$ is the weak limit of $\{\omega_j\}$ (or of an appropriate subsequence) or the first moment of ν, and $\hat{\psi} = G\hat{\omega}$ is the corresponding stream function. Notice that we have used the compactness of the Green's operator G. It should be noticed that the energy of ν resides in the mean field $\hat{\omega}$ since $\hat{E}(\nu) = E(\hat{\omega}) = E(\omega_0) = E^0$; the fluctuations do not contribute to the energy. On the other hand, the microscopic fluctuations do contribute to the entropy integrals since in general it is not true that $\hat{F}_f(\nu)$ has the same value as $F_f(\hat{\omega})$. We might say, therefore, that entropy is not conserved on a macroscopic scale, as part of it is lost to the infinitesimal-scale fluctuations of the vorticity.

We have demonstrated that if the parametrized measure ν is generated by a sequence of vorticity functions ω_j arising from a solution of the Euler equations, then it must satisfy the above energy and entropy constraints. Insofar as this is the only tangible information available about the possible long-time weak limit parametrized measures, we shall take as our admissible class of measures the set

$$\mathcal{A} = \left\{\nu = \{\nu_x\}_{x\in\Omega} : \operatorname{supp}(\nu_x) \subset [-r,r], \hat{E}(\nu) = E^0, \hat{F}_f(\nu) = F_f^0 \text{ for all } f\right\}.$$

We now seek to determine those elements in $\mathcal{A}$ that are in some sense most probable, and therefore the most likely to be observed as long-time equilibrium states of the Euler system. This task is accomplished through the introduction of the Kullback entropy functional

$$K_\pi(\nu) = -\int_\Omega \int_{[-r,r]} \log \frac{d\nu_x}{d\pi_0}\, d\nu_x(y)\, dx$$

if ν_x is absolutely continuous with respect to π_0. Otherwise it is taken to be $-\infty$. Here π_0 is a fixed probability measure on $[-r,r]$ and $\pi = dx \otimes \pi_0$ is a spatially homogeneous probability measure. The functional K is well known from information theory and statistical physics. As an integral in y it is a measure of the logarithm of the number of microscopic vorticity fields ω corresponding to the macrostate ν. The functional $I = -K$ is a measure of the statistical distance from ν to the homogeneous parametrized measure π. Thus if ν maximizes K over the admissible class $\mathcal{A}$, then ν minimizes the distance to π and ν is also most probable in the sense that it corresponds to the largest number of microstates ω. It is clear that the choice of the reference measure π is important. It has been argued that π_0 should be chosen to be the probability measure $(1/|\Omega|)\pi_\omega$, where π_ω is the vorticity distribution function defined by

$$\langle \pi_\omega, f\rangle = \int_\Omega f(\omega)\, dx.$$

This distribution function is conserved by the Eulerian flow, because the entropy integrals are invariant. The measure π then represents the most mixed, or most random macrostate. It satisfies the entropy constraints, but not the constraint on the energy. With this choice of π, the most probable parametrized measure consistent with both of these constraints is determined as a solution of the maximum entropy principle

$$K_\pi(\nu) \to \max \qquad \text{subject to } \nu \in \mathcal{A}.$$

While we have attempted to motivate the maximum entropy principle as an intutitively appealing procedure for selecting the most probable admissible parametrized measures, its rigorous justification rests upon methods from statistical mechanics and the theory of large deviations. These developments are beyond the scope of this text. We merely wish to point out that the set of solutions of the maximum entropy principle, $\mathcal{A}^*$, satisfies a natural concentration property, which roughly states that an overwhelming majority of the measures in the admissible class $\mathcal{A}$ concentrate about that subset of solutions. In particular, any parametrized measure that is generated by a sequence of vorticity fields corresponding to a solution of the Euler equations concentrates about $\mathcal{A}^*$.

6. Bibliographical remarks

References for Sections 2 and 3 are basic works on parametrized measures, the calculus of variations and optimization; these have already been mentioned in Chapter 1. Sections 4 and 5 are, however, more specific. An important subject from the point of view of applications not included in this chapter where weak convergence and homogenization play also a fundamental role is the theory of composites. A few references on this topic are [11], [146], [197], [235], [236], [237].

The main sources for the optimal design problem of Section 4 are [49], [50], [51], [198]. Numerical experiments are recorded in [71], [72]. The optimal relaxation as well as the general approach in terms of parametrized measure as it has been explained in this chapter can be found in [245]. [246] and [301] contain the basic results on H-convergence used in this problem.

A more detailed discussion of the statistical approach in terms of parametrized measures of turbulence as well as justification for some of our remarks in Section 5 can be studied in [28], [48], [119], [165], [177], [178], [180], [234], [273], [311].

Chapter 3
The Calculus of Variations under Convexity Assumptions

1. Introduction

The central focus of the calculus of variations is the functional

$$I(u) = \int_\Omega \varphi(x, u(x), \nabla u(x))\, dx,$$

where the integrand φ explicitly depends upon the gradient variable ∇u. Ω is assumed to be an open, regular, bounded domain of $\mathbf{R}^N$. The admissible functions $u : \Omega \to \mathbf{R}^m$ belong to some reflexive Sobolev space and they may satisfy some other restriction like having the boundary values prescribed. The integrand $\varphi : \Omega \times \mathbf{R}^m \times \mathbf{M}^{m\times N} \to \mathbf{R}^*$ is assumed to be a Carathéodory function. By this we simply mean that φ is measurable on the x variable and continuous with respect to u and ∇u. We may eventually let φ take on the value $+\infty$ as indicated by $\mathbf{R}^* = \mathbf{R} \cup \{+\infty\}$.

We devote the present chapter to proving results in the spirit of Theorem 1.1 for this type of functionals. The main difficulty is the weak lower semicontinuity property. We want to understand the conditions on φ that ensure this important property. This will take us to the quasiconvexity condition for φ, so that gradient parametrized measures will also play a crucial role in the analysis that follows. Since the quasiconvexity condition, except for the scalar case, is hard to grasp we look for sufficient conditions for quasiconvexity. Polyconvexity is then introduced as the main source of quasiconvex functions that are not convex. Our analysis does not pretend in any way to be complete in this regard. Having in mind applications of existence theorems for polyconvex integrands, we discuss very briefly three-dimensional elasticity in Section 5. Finally we explore how the fact of being a minimizing sequence for some functional provides further information that can be used to derive weak and strong convergence results and representation formulas in terms of gradient parametrized measures. Remember that a $W^{1,p}$-parametrized measure is the parametrized measure associated to a bounded sequence of gradients in $L^p(\Omega)$.

The proofs in this chapter are based on the results stated in Chapter 1, Sections 4 and 5. The complete proofs of those are contained in Chapters 6, 7 and 8. Because the space $W^{1,1}(\Omega)$ is not reflexive, the case $p=1$ is very special. Even though some of the conclusions in this chapter are valid for $p=1$, or may be restated in some way so that they become true, we consistently avoid this delicate case. We take $1<p<\infty$ throughout this chapter unless explicitly stated otherwise.

2. Weak lower semicontinuity

We start by giving the proof of a very general weak lower semicontinuity result for functionals I of the type described in the Introduction. We first consider the integrand φ depending on the gradient variable alone and move on to the case of full generality.

Theorem 3.1 *Let φ be a continuous function defined over matrices, bounded from below. Let $\{u_j\}$ be a sequence of $W^{1,p}(\Omega)$-functions converging weakly in $W^{1,p}(\Omega)$ to u. Let $\nu=\{\nu_x\}_{x\in\Omega}$ be the parametrized measure associated to $\{\nabla u_j\}$ (or possibly to a subsequence), so that*

$$\nabla u(x)=\int_{\mathbf{M}^{m\times N}} A\,d\nu_x(A), \qquad \text{a.e. } x\in\Omega.$$

If $\liminf_{j\to\infty}\int_\Omega \varphi(\nabla u_j)\,dx<\infty$, then

$$\int_E \varphi(\nabla u)\,dx\le \liminf_{j\to\infty}\int_E \varphi(\nabla u_j)\,dx,$$

for all measurable $E\subset\Omega$, if and only if

$$\varphi(\nabla u(x))\le \int_{\mathbf{M}^{m\times N}} \varphi(A)\,d\nu_x(A), \qquad \text{a.e. } x\in\Omega. \tag{3–1}$$

Proof. Recalling Theorem 6.11 and assuming Jensen's inequality (3–1) we have

$$\begin{aligned}\liminf_{j\to\infty}\int_E \varphi(\nabla u_j)\,dx &\ge \int_E\int_{\mathbf{M}^{m\times N}} \varphi(A)\,d\nu_x(A)\,dx\\ &\ge \int_E \varphi(\nabla u(x))\,dx.\end{aligned}$$

Conversely, if $\{\varphi(\nabla u_j)\}$ is a uniformly bounded sequence in $L^1(\Omega)$, by Chacon's biting lemma there exists a nested sequence of measurable sets Ω_k whose measures tend to 0, and such that

$$\begin{aligned}\varphi(\nabla u_j) &\rightharpoonup \overline{\varphi} \quad \text{in } L^1(\Omega\setminus\Omega_k), \text{ for all } k,\\ \overline{\varphi}(x) &= \int_{\mathbf{M}^{m\times N}} \varphi(A)\,d\nu_x(A).\end{aligned}$$

By assumption

$$\begin{aligned}\int_{E\setminus\Omega_k} \varphi(\nabla u)\,dx &\le \lim_{j\to\infty} \int_{E\setminus\Omega_k} \varphi(\nabla u_j)\,dx \\ &= \int_{E\setminus\Omega_k} \int_{\mathbf{M}^{m\times N}} \varphi(A)\,d\nu_x(A)\,dx \\ &= \int_{E\setminus\Omega_k} \overline{\varphi}(x)\,dx.\end{aligned}$$

By Theorem 6.11, $\overline{\varphi} \in L^1(\Omega)$ and by hypothesis and the weak lower semicontinuity property $\varphi(\nabla u) \in L^1(\Omega)$. Therefore, since $|\Omega_k| \to 0$, taking limits in k,

$$\int_E \varphi(\nabla u)\,dx \le \int_E \int_{\mathbf{M}^{m\times N}} \varphi(A)\,d\nu_x(A)\,dx.$$

The arbitrariness of E yields (3–1). ■

The proof of this theorem in the general case in which φ depends also upon x and u is not more difficult.

Theorem 3.2 *Let $\varphi : \Omega \times \mathbf{R}^m \times \mathbf{M}^{m\times N} \to \mathbf{R}^*$ be non-negative (or bounded from below). Let $\{u_j\}$ be a sequence of $W^{1,p}(\Omega)$-functions converging weakly in $W^{1,p}(\Omega)$ and strongly in $L^p(\Omega)$ to u. Let $\nu = \{\nu_x\}_{x\in\Omega}$ be the parametrized measure corresponding to $\{\nabla u_j\}$ so that*

$$\nabla u(x) = \int_{\mathbf{M}^{m\times N}} A\,d\nu_x(A).$$

If $\{\varphi(x, u_j(x), \nabla u_j(x))\}$ is a uniformly bounded sequence in $L^1(\Omega)$, then

$$\int_E \varphi(x, u, \nabla u)\,dx \le \liminf_{j\to\infty} \int_E \varphi(x, u_j, \nabla u_j)\,dx, \quad \text{for all } E \subset \Omega,$$

if and only if

$$\varphi(x, u(x), \nabla u(x)) \le \int_{\mathbf{M}^{m\times N}} \varphi(x, u(x), A)\,d\nu_x(A), \quad \text{a.e. } x \in \Omega. \tag{3–2}$$

For the proof simply notice that by Proposition 6.13 and the comments after it, the parametrized measure associated to $\{(u_j, \nabla u_j)\}$ is $\left\{\delta_{u(x)} \otimes \nu_x\right\}_{x\in\Omega}$ and follow the steps of the previous proof.

The two theorems above are concerned with a particular sequence of functions and a particular energy density so that if inequality (3–2) is verified for a given parametrized measure and energy density φ, then the weak lower semicontinuity result will hold for any sequence generating that parametrized measure. If we would like to have a general condition for every possible sequence in $W^{1,p}(\Omega)$ converging weakly, we would obtain the following "convexity" condition on the energy density φ:

$$\int_{\mathbf{M}^{m\times N}} \varphi(x,u,A)\,d\nu(A) \geq \varphi\left(x,u,\int_{\mathbf{M}^{m\times N}} A\,d\nu(A)\right), \tag{3–3}$$

for all $(x,u) \in \Omega \times \mathbf{R}^m$ and for all ν, a homogeneous $W^{1,p}$-parametrized measure. Notice that by the localization principle Theorem 8.4 each ν_x can be regarded as a homogeneous $W^{1,p}$-parametrized measure by itself.

This convexity condition (3–3) takes us one step further than quasiconvexity or $W^{1,p}$-quasiconvexity. Let us assume that φ is bounded from below and depends upon the gradient variable alone. Take $u \in W^{1,p}(\Omega)$ such that $u - u_Y \in W_0^{1,p}(\Omega)$ where $u_Y(x) = Y\,x$ and Y is any fixed matrix. We can consider the average of $\delta_{\nabla u(x)}$, ν_u, which is a homogeneous $W^{1,p}$-parametrized measure with underlying deformation Y according to Theorem 8.1. For ν_u, (3–3) reduces to

$$\varphi(Y) \leq \frac{1}{|\Omega|}\int_\Omega \varphi(\nabla u(x))\,dx,$$

which is the usual $W^{1,p}$-quasiconvexity condition. In (3–3), however, we are demanding the inequality for "all" homogeneous $W^{1,p}$-parametrized measures and not just for those coming from single functions. In so doing, we are requiring that concentration effects do not have any influence on the quasiconvexity inequality when sequences are considered. In order to better explain this issue, suppose we have a bounded sequence $\{u_j\} \subset W^{1,p}(\Omega)$ with $u_j - u_Y \in W_0^{1,p}(\Omega)$. By the quasiconvexity condition

$$\varphi(Y) \leq \liminf_{j\to\infty} \frac{1}{|\Omega|}\int_\Omega \varphi(\nabla u_j(x))\,dx.$$

On the other hand if $\nu = \{\nu_x\}_{x\in\Omega}$ is the parametrized measure associated to $\{\nabla u_j\}$, since we have affine boundary conditions, we can consider the averaged, homogeneous $W^{1,p}$-parametrized measure, $\overline{\nu}$. If the sequence $\{\varphi(\nabla u_j)\}$ does not develop concentrations, so that it is a weak convergent sequence in $L^1(\Omega)$, then (3–3) is a corollary of the usual $W^{1,p}$-quasiconvexity condition because in this case for a suitable subsequence

$$\int_{\mathbf{M}^{m\times N}} \varphi(A)\,d\overline{\nu}(A) = \lim_{j\to\infty} \frac{1}{|\Omega|}\int_\Omega \varphi(\nabla u_j(x))\,dx,$$

and (3–3) holds. If, on the other hand, $\{\varphi(\nabla u_j)\}$ develops concentrations, then

$$\int_{\mathbf{M}^{m\times N}} \varphi(A)\,d\overline{\nu}(A) < \lim_{j\to\infty} \frac{1}{|\Omega|}\int_\Omega \varphi(\nabla u_j(x))\,dx,$$

but still (3–3) requires

$$\varphi(Y) \leq \int_{\mathbf{M}^{m\times N}} \varphi(A)\,d\overline{\nu}(A).$$

Theorem 3.3 *Consider the functional*

$$I(u) = \int_\Omega \varphi(x, u(x), \nabla u(x))\, dx \tag{3–4}$$

defined on $W^{1,p}(\Omega)$ *where* φ *is non-negative and can eventually take on the value* $+\infty$. I *is weak lower semicontinuous in* $W^{1,p}(\Omega)$ *over arbitrary measurable sets if and only if* φ *is "convex" in the sense just described.*

Proof. i) Necessity. Let $(x_0, u_0) \in \Omega \times \mathbf{R}^m$ be given and ν, a homogeneous $W^{1,p}$-parametrized measure generated by $\{\nabla u_j\}$. Set $Y = \int_{\mathbf{M}^{m\times N}} A\, d\nu$ and $u(x) = Y(x - x_0) + u_0$. By adding suitable constants to u_j we may assume without loss of generality that $\{u_j\}$ converges weakly in $W^{1,p}(\Omega)$ and strongly in $L^p(\Omega)$ to u. By Theorem 3.2,

$$\int_{\mathbf{M}^{m\times N}} \varphi(x_0, u_0, A)\, d\nu(A) \geq \varphi(x_0, u_0, Y).$$

ii) Sufficiency. Let $\{u_j\}$ be any sequence converging weakly in $W^{1,p}(\Omega)$ and strongly in $L^p(\Omega)$ to u and let $\nu = \{\nu_x\}_{x\in\Omega}$ be the parametrized measure associated to $\{\nabla u_j\}$. By the localization property (Theorem 8.4) we know that each individual ν_x is a homogeneous $W^{1,p}$-parametrized measure, therefore by the "convexity" property (3–3) for $u = u(x)$ and $\nu = \nu_x$

$$\int_{\mathbf{M}^{m\times N}} \varphi(x, u(x), A)\, d\nu_x(A) \geq \varphi(x, u(x), \nabla u(x)),$$

for a.e. $x \in \Omega$. By Theorem 3.2, the weak lower semicontinuity property holds over arbitrary measurable sets. ∎

We describe below three situations in which this "convexity" condition can be translated into a more familiar convexity condition.

Theorem 3.4 *Let* $\varphi : \Omega \times \mathbf{R}^m \times \mathbf{M}^{m\times N} \to \mathbf{R}$ *be a finite, continuous energy density. The functional* $I(u)$ *in (3–4) is lower semicontinuous with respect to weak convergence in* $W^{1,\infty}(\Omega)$ *over arbitrary measurable sets if and only if* $\varphi(x_0, u_0, \cdot)$ *is quasiconvex for any* $x_0 \in \Omega$ *and* $u_0 \in \mathbf{R}^m$.

Proof. Let Y be any matrix and $u \in W^{1,\infty}(\Omega)$ such that $u - u_Y \in W_0^{1,\infty}(\Omega)$. The averaged parametrized measure, ν, corresponding to $\delta_{\nabla u(x)}$ is a $W^{1,\infty}$-parametrized measure. Therefore

$$\begin{aligned}\frac{1}{|\Omega|}\int_\Omega \varphi(x_0, u_0, \nabla u(x))\, dx = &\int_{\mathbf{M}^{m\times N}} \varphi(x_0, u_0, A)\, d\nu(A)\\ &\geq \varphi(x_0, u_0, Y),\end{aligned}$$

where

$$Y = \int_{\mathbf{M}^{m\times N}} A\, d\nu(A) = \frac{1}{|\Omega|}\int_\Omega \nabla u(x)\, dx,$$

and $\varphi(x_0, u_0, \cdot)$ is quasiconvex.

Conversely, let us assume that $\varphi(x_0, u_0, \cdot)$ is quasiconvex. If ν is any $W^{1,\infty}$-parametrized measure, there exists a sequence of functions, $\{u_j\}$, in $W^{1,\infty}(\Omega)$ whose gradients generate ν and $u_j - u_Y \in W_0^{1,\infty}(\Omega)$ for all j. In this case (because of the uniform boundedness of ∇u_j) $\{\varphi(x_0, u_0, \nabla u_j(x))\}$ is weakly convergent in $L^1(\Omega)$ and therefore

$$\int_{\mathbf{M}^{m\times N}} \varphi(x_0, u_0, A)\, d\nu(A) = \lim_{j\to\infty} \frac{1}{|\Omega|} \int_\Omega \varphi(x_0, u_0, \nabla u_j(x))\, dx$$
$$\geq \varphi(x_0, u_0, Y),$$

by the quasiconvexity property. ∎

Theorem 3.5 *Let $\varphi : \Omega \times \mathbf{R}^m \times \mathbf{M}^{m\times N} \to \mathbf{R}$ be a finite, continuous energy density such that*

$$0 \leq \varphi(x, u, A) \leq h(x, u)(1 + |A|^p)$$

where $h \in L^\infty_{loc}(\Omega \times \mathbf{R}^m)$ is continuous in u and $1 < p < \infty$. The functional $I(u)$ in (3–4) is lower semicontinuous with respect to weak convergence in $W^{1,p}(\Omega)$ over arbitrary measurable sets if and only if $\varphi(x_0, u_0, \cdot)$ is quasiconvex for any $x_0 \in \Omega$ and $u_0 \in \mathbf{R}^m$.

Proof. By Theorem 3.3 we have to check again that under these assumptions quasiconvexity is equivalent to the "convexity" condition in terms of $W^{1,p}$-parametrized measures.

Assume that $\varphi(x_0, u_0, \cdot)$ is quasiconvex for a.e. $x_0 \in \Omega$ and every $u_0 \in \mathbf{R}^m$. Let ν be a probability measure generated by a sequence of gradients, $\{\nabla u_j\}$, where $u_j \in W^{1,p}(\Omega)$ and $u_j - u_Y \in W_0^{1,p}(\Omega)$ for all j. By Corollary 8.15, $\{|\nabla u_j|^p\}$ is equiintegrable in $L^1(\Omega)$, and thus $\{\varphi(x_0, u_0, \nabla u_j(x))\}$ converges weakly in $L^1(\Omega)$ (notice that $h(x_0, u_0)$ is finite for a.e. $x_0 \in \Omega$ and every $u_0 \in \mathbf{R}^m$). Finally

$$\varphi(x_0, u_0, Y) \leq \liminf_{j\to\infty} \frac{1}{|\Omega|} \int_\Omega \varphi(x_0, u_0, \nabla u_j(x))\, dx$$
$$= \int_{\mathbf{M}^{m\times N}} \varphi(x_0, u_0, A)\, d\nu(A).$$

The converse follows immediately from the same argument given in the previous theorem. ∎

If we consider scalar problems where $m = 1$ or $N = 1$ the "convexity" condition is nothing but the classical Jensen's inequality. Indeed, in either of these two cases quasiconvexity, and therefore rank-one convexity, reduces to convexity since all directions on $\mathbf{M}^{m\times N}$ are of rank-one. Thus the weak lower semicontinuity property is valid if and only if the integrand is convex in the usual sense.

Theorem 3.6 *Let $\varphi : \Omega \times \mathbf{R} \times \mathbf{M}^{1\times N} \to \mathbf{R}^*$ (respectively $\varphi : \Omega \times \mathbf{R}^m \times \mathbf{M}^{m\times 1} \to \mathbf{R}^*$, $\Omega \subset \mathbf{R}$) be a Carathéodory function. The functional $I(u)$ where the competing functions are scalar valued $u : \Omega \subset \mathbf{R}^N \to \mathbf{R}$ (respectively $u : \Omega \subset \mathbf{R} \to \mathbf{R}^m$), is lower semicontinuous with respect to weak convergence in $W^{1,p}(\Omega)$, $1 \le p \le \infty$, over arbitrary measurable sets if and only if $\varphi(x_0, u_0, \cdot)$ is convex for a.e. $x_0 \in \Omega$ and every $u_0 \in \mathbf{R}$ (respectively $u_0 \in \mathbf{R}^m$).*

If we do not assume any growth on φ in the vector case one would desire that the "convexity" condition be equivalent to $W^{1,p}$-quasiconvexity, or at least that $W^{1,p}$-quasiconvexity is the condition necessary and sufficient in order to have weak lower semicontinuity over regular, open subsets of Ω. Either of these two questions is still open.

3. Existence theorems

Bringing together these weak lower semicontinuity results and the abstract setting for the direct method discussed in Chapter 1, one gets immediately the following existence theorems. The first one is concerned with the scalar case when $\Omega \subset \mathbf{R}$ or the competing functions u are scalar-valued. The other two correspond to the fully vector case in two situations: under growth restrictions on the energy density and without these conditions. In all three theorems, $\varphi : \Omega \times \mathbf{R}^m \times \mathbf{M}^{m\times N} \to \mathbf{R}^*$ will stand as usual for a non-negative energy density. Either of the cases $m = 1$ or $N = 1$ correspond to the scalar case. The functional we are considering is again

$$I(u) = \int_\Omega \varphi(x, u(x), \nabla u(x))\, dx,$$

and the variational principle for which we would like to prove existence theorems is

$$\inf\left\{ I(u) : u \in W^{1,p}(\Omega), u - u_0 \in W_0^{1,p}(\Omega)\right\}, \tag{P}$$

for some prescribed $u_0 \in W^{1,p}(\Omega)$ such that $I(u_0) < \infty$. In this situation the set of admissible functions is $\mathcal{A} = \left\{u \in W^{1,p}(\Omega) : u - u_0 \in W_0^{1,p}(\Omega)\right\}$ which is weakly closed in $W^{1,p}(\Omega)$. The condition on the reflexivity of the underlying Banach space prevents us from taking $p = 1$, so that $1 < p < \infty$ as usual. The case $p = \infty$ is also discarded because bounds in the $W^{1,\infty}$-norm for a minimizing sequence cannot be obtained from coercivity bounds on the energy density.

Theorem 3.7 *Let $m = 1$ or $N = 1$. Let φ be a Carathéodory function, convex on the gradient variable and*

$$\max(0, c\,|A|^p - 1) \le \varphi(x, u, A),$$

for a.e. $x \in \Omega$ and all $u \in \mathbf{R}^m$, where $c > 0$ is a constant. The variational principle (P) admits at least a minimizer.

The coercivity property for the functional is obtained from the lower bound on φ and Poincaré's inequality bearing in mind that $u - u_0 \in W_0^{1,p}(\Omega)$ for all competing functions u. If, moreover, φ is strictly convex, the minimizer is unique. The argument for this uniqueness is elementary.

Theorem 3.8 *Let $m > 1$ and $N > 1$. If φ is a Carathéodory function, quasiconvex on the gradient variable and*

$$\max(0, c\,|A|^p - 1) \leq \varphi(x, u, A) \leq h(x, u)(1 + |A|^p),$$

where $h \in L^\infty_{loc}(\Omega \times \mathbf{R}^m)$ and $c > 0$, the variational principle (P) admits at least a minimizer.

If we drop the upper bound on φ, the quasiconvexity condition should be strengthened to the "convexity" condition (3–3) in terms of homogeneous $W^{1,p}$-parametrized measures.

Theorem 3.9 *Let $m > 1$ and $N > 1$. If φ is a Carathéodory function such that*

$$\max(0, c\,|A|^p - 1) \leq \varphi(x, u, A),$$

and

$$\int_{\mathbf{M}^{m\times N}} \varphi(x, u, A)\, d\nu(A) \geq \varphi\left(x, u, \int_{\mathbf{M}^{m\times N}} A\, d\nu(A)\right),$$

for all $(x, u) \in \Omega \times \mathbf{R}^m$ and for all ν, homogeneous $W^{1,p}$-parametrized measures, the variational principle (P) admits at least a minimizer.

At this point, it should be clear how important convexity conditions are in order to have weak lower semicontinuity in different contexts. In the vector case, we have sufficiently emphasized that the quasiconvexity condition is the appropriate notion. Unfortunately, in practice, it is almost impossible to decide whether an explicit function is quasiconvex and almost impossible to compute the quasiconvexification even in the simplest, nontrivial cases. The difficulty is related to the essentially nonlocal nature of the definition of quasiconvexity and our inability to understand how gradients are built. Since the pioneering work of Morrey, there have been several attempts to further understand the notion of quasiconvexity, and in particular to decide whether quasiconvexity is equivalent to rank-one convexity. As we now know, rank-one convexity is weaker than quasiconvexity. We will show this in Chapter 9. There are some situations, though, where by restricting further the form of the function one can show that rank-one convexity is indeed equivalent to quasiconvexity. The most important example is when φ is quadratic in the gradient variable. A different notion of convexity, called polyconvexity, is introduced as a convenient, sufficient condition for quasiconvexity, thus providing the main source for quasiconvex functions which are not convex. We turn to study this new type of convexity.

4. Polyconvexity

To explain this new convexity, one needs to identify the weak continuous functionals: the continuous functions φ defined on matrices, such that both φ and $-\varphi$ are quasiconvex. In this case we would have that if $u_j \overset{*}{\rightharpoonup} u$ in $W^{1,\infty}(\Omega)$, then

$$\lim_{j\to\infty} \int_\Omega \varphi(\nabla u_j)\,dx = \int_\Omega \varphi(\nabla u)\,dx.$$

Under suitable upper estimates on φ, we could have the same weak continuity result for weak convergence in $W^{1,p}(\Omega)$.

A first idea is that since quasiconvexity implies rank-one convexity, φ and $-\varphi$ will have to be rank-one convex, i.e. the function $\varphi(A+tB)$ must be affine in t provided B has rank one. The functions that enjoy this property are called quasiaffine functions and they are identified in the following theorem which we do not prove here.

Theorem 3.10 *The minors of a gradient matrix are the only quasiaffine functions.*

In the case of 2×2 matrices, there are five minors: the four entries (the 1×1 minors) and the determinant. For 3×3 matrices, we have nineteen minors: the nine entries, nine 2×2 minors and the unique 3×3 minor, the determinant. And so on.

The fact that these minors are quasiaffine is easy to check. As a matter of fact, it is not hard to convince oneself that for one of these minors, φ, the function of t $\varphi(A + tB)$ is a polynomial of degree rank(B). That these are the only quasiaffine functions requires more work and a little bit of algebraic manipulation. See the last section for more references.

We know where to look for the functions which give rise to weak continuous functionals. By an abuse of language, we refer to those functions as weak continuous functions. In order to be more precise, we need to introduce some notation.

Let A stand for any matrix in $\mathbf{M}^{m\times N}$ and A' for any square submatrix of A. $\operatorname{adj}(A')$ is the matrix of cofactors of A' so that

$$A' \operatorname{adj}(A') = \det(A')\,\mathbf{1},$$

where $\mathbf{1}$ is the identity matrix in the appropriate dimension. If A' is nonsingular, then

$$\operatorname{adj}(A') = \det(A')(A')^{-1}.$$

Let $r = \max\{m, N\}$ and $p > r$.

Theorem 3.11 *If $u_j \rightharpoonup u$ in $W^{1,p}(\Omega)$ then $\det(\nabla u_j)' \rightharpoonup \det(\nabla u)'$ in $L^{p/r}(\Omega)$.*

Proof. We divide the proof in several steps.

Step 1. Let $v \in W^{1,p}(\Omega)$. We claim that $\operatorname{div}(\operatorname{adj}(\nabla v)') = 0$ in the sense of distributions. Assume first that v is actually smooth. Based on the equality of the mixed partial derivatives, we find indeed that $\operatorname{div}(\operatorname{adj}(\nabla v)') = 0$. For a general $v \in W^{1,p}(\Omega)$, take a sequence of smooth functions, $\{v_j\}$, converging strongly to v in $W^{1,p}(\Omega)$. $\operatorname{adj}(\nabla v_j)'$ converges strongly to $\operatorname{adj}(\nabla v)'$ in $L^{p/r}(\Omega)$ (using Hölder's inequality) because the terms in $\operatorname{adj}(\nabla v_j)'$ are products of at most $r-1$ factors. For a smooth test function ψ

$$\int_\Omega \operatorname{adj}(\nabla v_j)' \nabla \psi \, dx = 0,$$

for all j. By the strong convergence just pointed out

$$\int_\Omega \operatorname{adj}(\nabla v)' \nabla \psi \, dx = 0,$$

and this is our claim.

Step 2. As a consequence of Step 1, we obtain that

$$\operatorname{div}'(u \operatorname{adj}(\nabla u)') = (\nabla u)' \operatorname{adj}(\nabla u)' = \det(\nabla u)'$$

as distributions where div' means divergence with respect to the variables involved in the submatrix A' that has been determined previously. By induction, let us assume that $\operatorname{adj}(\nabla u_j)' \rightharpoonup \operatorname{adj}(\nabla u)'$ in $L^{p/r}(\Omega)$. If ψ is a smooth test function, then

$$\int_\Omega \det(\nabla u_j)' \psi \, dx = \int_\Omega u_j \operatorname{adj}(\nabla u_j)' \nabla \psi \, dx. \tag{3–5}$$

But $\{u_j\}$ converges strongly to u in $L^\infty(\Omega)$ by the Compactness Theorem for Sobolev functions ($p > N$). Hence the limit in (3–5) is

$$\int_\Omega u \operatorname{adj}(\nabla u)' \nabla \psi \, dx = \int_\Omega \det(\nabla u)' \psi \, dx,$$

and $\det(\nabla u_j)'$ converges weakly in the sense of distributions to $\det(\nabla u)'$.

Step 3. Conclusion. We also have a uniform bound on $\det(\nabla u_j)'$ in $L^{p/r}(\Omega)$ because terms in $\det(\nabla u_j)'$ have less than r factors. Since $p/r > 1$, at least for a subsequence (not relabeled) $\det(\nabla u_j)'$ converges weakly in $L^{p/r}(\Omega)$. By Step 2 and the uniqueness of the limit we conclude that in fact $\det(\nabla u_j)'$ converges weakly in $L^{p/r}(\Omega)$ to $\det(\nabla u)'$. ∎

This result holds for $r = \min\{m, N\}$ as well. The proof is the same. It only requires a more careful analysis of exponents.

Let $M(A)$ represent the vector of all possible minors of any dimension of A considered in some order. A continuous function $\varphi : \mathbf{M}^{m\times N} \to \mathbf{R}^*$ is called polyconvex if it can be rewriten as $g(M(A))$ where g is a convex function of all its arguments (convex in the usual sense). The most important property of polyconvex functions is that they are quasiconvex.

Proposition 3.12 *Let φ be a polyconvex function. For $p \geq r$, φ satisfies Jensen's inequality (3–3) for any homogeneous $W^{1,p}$-parametrized measure.*

Proof. The proof is simple. Assume that we have a uniformly bounded sequence in $W^{1,p}(\Omega)$, $\{u_j\}$, generating a homogeneous parametrized measure ν with first moment Y: $Y = \int A\, d\nu(A)$. Without loss of generality we may well assume that $\{|\nabla u_j|^p\}$ is equiintegrable according to Lemma 8.15. Since $\{u_j\}$ converges weakly to u_Y, affine, by the previously established weak convergence for $p > r$,

$$M(\nabla u_j) \rightharpoonup M(Y) \quad \text{in } L^{p/r}(\Omega).$$

This is also true for $p = r$ by the assumed equiintegrability and the fact that minors of any order are bounded above by the power corresponding to its order,

$$|M(A)| \leq C(1 + |A|^r).$$

Because of this weak convergence the representation in terms of ν is valid

$$M(\nabla u_j) \rightharpoonup \int_{\mathbf{M}^{m\times N}} M(A)\, d\nu(A).$$

Hence

$$\int_{\mathbf{M}^{m\times N}} M(A)\, d\nu(A) = M(Y) = M\left(\int_{\mathbf{M}^{m\times N}} A\, d\nu(A)\right).$$

Since g is convex, by Jensen's inequality,

$$\begin{aligned}\int_{\mathbf{M}^{m\times N}} \varphi(A)\, d\nu(A) =& \int_{\mathbf{M}^{m\times N}} g(M(A))\, d\nu(A)\\ \geq& g\left(\int_{\mathbf{M}^{m\times N}} M(A)\, d\nu(A)\right)\\ =& g(M(Y))\\ =& \varphi(Y).\end{aligned}$$

■

As a consequence, any polyconvex function is quasiconvex. We can now write down many non-trivial examples of quasiconvex functions. For example, in the case $m = N$, any convex function of the determinant is quasiconvex (notice that the determinant itself is not convex). One particularly important example is the jacobian: $\varphi(A) = |\det A|$. Because of the upper bound on the determinant

$$0 \leq |\det A| \leq C(1 + |A|^N),$$

we conclude that the jacobian is $W^{1,p}$-quasiconvex for $p \geq N$.

If we are willing to accept also dependence of φ on x and u, polyconvexity is defined in the same way for a.e. $x \in \Omega$ and all $u \in \mathbf{R}^m$. We have the following existence theorem for polyconvex integrands which is a corollary of Theorem 3.9.

Theorem 3.13 *If $\varphi : \Omega \times \mathbf{R}^m \times \mathbf{M}^{m\times N} \to \mathbf{R}^*$ is nonnegative and polyconvex, and for $p \geq r$ ($r = \max\{m, N\}$)*

$$c\,|A|^p - 1 \leq \varphi(x, u, A), \quad c > 0,$$

for all $A \in \mathbf{M}^{m\times N}$, a.e. $x \in \Omega$ and all $u \in \mathbf{R}^m$, the variational principle (P) with integrand φ admits at least a minimizer.

More precise statements about polyconvexity and existence theorems can be found in the references (see Section 7).

In the last few years a fairly large amount of work has been done trying to relate and understand all these different notions of convexity. In particular, counterexamples have been produced to show that quasiconvexity is strictly stronger than polyconvexity. We refer the reader to the bibliography.

5. A brief account of nonlinear elasticity

This section presumes to be only a very short and basic review of the mathematical theory of nonlinear elasticity. The aim is to emphasize the importance of variational principles for the vector case, and the crucial role that polyconvexity plays. For the sake of brevity, we will not make precise statements.

There are materials in nature whose equilibrium configurations in various enviroments can be understood through an energy minimization principle. The material seeks the minimum energy available to it under the prescribed conditions. In this sense we identify minimizers of the energy functional with equilibrium states. The possible deformations that a material may undergo are described mathematically by means of a vector function $u : \Omega \to \mathbf{R}^3$ where $\Omega \subset \mathbf{R}^3$ is the reference configuration with respect to which we consider all deformations. The gradient ∇u is referred to as the deformation gradient and intuitively represents the local deformation or strain around each point x in the reference configuration. For the type of materials we are interested in, we assume the existence of a continuous stored energy density φ defined on 3×3 matrices so that the free energy associated to a particular deformation is measured by the integral

$$\int_\Omega \varphi(\nabla u(x))\,dx.$$

From the physical point of view, the energy density φ must comply with several restrictions. For instance, it should be material frame-indifferent. We must require

$$\varphi(F) = \varphi(QF),$$

for all proper rotations Q in space (by proper we mean positive determinant rotations). Moreover, φ must also satisfy the condition

$$\varphi(F) \to +\infty \text{ if } \det F \to 0, \quad \varphi(F) = +\infty \text{ if } \det F \leq 0,$$

to reflect the fact that infinite energy is associated with "extreme" deformations trying to collapse some volume into a plane or a line, although this condition is often relaxed. Further restrictions can be imposed depending on specific properties of the material under consideration. These constraints have deep implications concerning the structure of φ.

One of the most common situations consists in determining the equilibrium configurations of the material under prescribed boundary values. This is accomplished by determining the boundary values on $\partial\Omega$ that competing deformations should have. In this framework, equilibrium configurations will correspond to minimizers of the variational principle

$$\int_\Omega \varphi(\nabla u(x))\,dx, \quad u \in W^{1,p}(\Omega),\ u - u_0 \in W_0^{1,p}(\Omega).$$

We are faced with a variational problem of the type we have been discussing so far. The existence of such equilibrium states is closely connected to the "convexity" properties of the energy density φ. One striking consequence of the behavior of the energy density for minimizing deformations is that φ cannot be convex in the usual sense. This reasonable assumption rules out the possibility of having convex energy densities. The axiom of frame-indifference also has serious implications concerning the eigenvalues of the Cauchy stress tensor $\nabla u^T \nabla u$.

A crucial observation is that these difficulties are not present when considering polyconvex stored energy functions φ. An important class of polyconvex functions that appear as energy densities in nonlinear elasticity is

$$\varphi(F) = \sum_{i=1}^{t} a_i \operatorname{tr}(F^T F)^{\alpha_i/2} + \sum_{j=1}^{s} \operatorname{tr}(\operatorname{adj} F^T F)^{\beta_j/2} + g(\det F),$$

where tr stands for the trace of a matrix, s, t are positive integers, $a_i > 0$, $\alpha_i \geq 1, \beta_j \geq 1$, and g is a convex function. This function satisfies a coerciveness inequality as well,

$$\varphi(F) \geq \alpha \left(\|F\|^p + \|\operatorname{adj} F\|^q\right) + g(\det F),$$

where $\alpha > 0$, $p = \max \alpha_i$ and $q = \max \beta_j$. A material whose energy density is of the above type and satisfies the additional property $\lim_{\lambda\to 0} g(\lambda) = +\infty$ is called an Ogden material. Particular examples are:

1. Neo-Hookean materials:

$$\varphi(F) = a \|F\|^2 + g(\det F), \quad a > 0.$$

2. Mooney-Rivlin materials:

$$\varphi(F) = a \|F\|^2 + b \|\operatorname{adj} F\|^2 + g(\det F), \quad a > 0, b > 0.$$

For materials that admit this sort of stored energy density, the existence of equilibrium configurations can be easily established in the framework of the direct method of the calculus of variations. There are, however, some examples for which the energy density is not polyconvex. One such example is the St. Venant-Kirchhoff materials:

$$\varphi(F) = a(\mathrm{tr}E)^2 + b\ \mathrm{tr}E^2, \quad \mathbf{1} + 2E = F^T F.$$

6. Weak and strong convergence of minimizing sequences

Convergence of energies together with weak convergence enable us to obtain some useful weak convergence results in $L^1(\Omega)$. This fact has important consequences for representation formulas in terms of parametrized measures for minimizing sequences. The issue we address is how the fact that a particular sequence is minimizing for some functional may help in deriving weak convergence results.

Theorem 3.14 *Let φ be continuous, quasiconvex and satisfy*

$$0 \le \varphi(A) \le C(1 + |A|^p), \quad A \in \mathbf{M}^{m\times N}, p > 1.$$

Suppose that

$$u_j \overset{*}{\rightharpoonup} u \text{ in } W^{1,p}(\Omega),$$

$$\int_\Omega \varphi(\nabla u)\, dx = \lim_{j\to\infty} \int_\Omega \varphi(\nabla u_j)\, dx.$$

There exists a subsequence (not relabeled) such that

$$\varphi(\nabla u_j) \overset{*}{\rightharpoonup} \varphi(\nabla u) \text{ in } L^1(\Omega).$$

Proof. Assume, by contradiction, that $\{\varphi(\nabla u_j)\}$ is not equiintegrable. For some $\epsilon > 0$ and every $\delta > 0$, there exists $A_\delta \subset \Omega$ and an integer j_δ such that $|A_\delta| < \delta$ and

$$\int_{A_\delta} \varphi(\nabla u_{j_\delta})\, dx > \epsilon.$$

Since $\varphi(\nabla u)$ belongs to $L^1(\Omega)$, there is a $\delta_0 > 0$ such that if $|E| < \delta_0$, then

$$\int_E \varphi(\nabla u)\, dx < \epsilon. \tag{3–6}$$

Let us choose in particular $\delta_k = 2^{-k}\delta_0$. There exists a sequence A_k, $|A_k| < \delta_k$, and j_k such that

$$\int_{A_k} \varphi(\nabla u_{j_k})\, dx > \epsilon, \text{ for all } k.$$

Let $E = \bigcup_k A_k$, so that $|E| \leq \delta_0$ and (3–6) holds. Thus

$$\epsilon \leq \int_E \varphi(\nabla u_{j_k})\, dx = \int_\Omega \varphi(\nabla u_{j_k})\, dx - \int_{\Omega\setminus E} \varphi(\nabla u_{j_k})\, dx.$$

Letting $j_k \to \infty$, we have by weak lower semicontinuity and the convergence of energies that

$$\begin{aligned}\epsilon &\leq \int_\Omega \varphi(\nabla u)\, dx - \int_{\Omega\setminus E} \varphi(\nabla u)\, dx \\ &= \int_E \varphi(\nabla u)\, dx \\ &< \epsilon,\end{aligned}$$

a contradiction. ∎

The condition on the convergence of energies is automatic for minimizing sequences. The weak convergence of Theorem 3.14 together with the coercivity of φ can be used to obtain representation of weak limits through parametrized measures associated with minimizing sequences. Let X^p be the space of functions defined on matrices with growth of order less than p:

$$X^p = \left\{\varphi : \mathbf{M}^{m\times N} \to \mathbf{R},\ \text{continuous} : |\varphi(A)| \leq C\,(1 + |A|^p)\right\}.$$

Corollary 3.15 *Let φ be a quasiconvex function such that*

$$\max(0, c\,|A|^p - 1) \leq \varphi(A) \leq C(|A|^p + 1), \quad 1 \leq p < \infty, 0 < c \leq C.$$

Consider the variational principle

$$\inf\left\{\int_\Omega \varphi(\nabla u)\, dx : u \in W^{1,p}(\Omega), u - u_0 \in W_0^{1,p}(\Omega)\right\},$$

for $u_0 \in W^{1,p}(\Omega)$. If $\{u_j\}$ is minimizing, $u_j \rightharpoonup u$ (so that u is a minimizer) and $\nu = \{\nu_x\}_{x\in\Omega}$ is the parametrized measure corresponding to $\{\nabla u_j\}$, then

$$\varphi(\nabla u_j) \rightharpoonup \varphi(\nabla u) \quad \text{in } L^1(\Omega),$$

and

$$\psi(\nabla u_j) \rightharpoonup \overline{\psi} \quad \text{in } L^1(\Omega),$$

for any $\psi \in X^p$ where, as usual,

$$\overline{\psi}(x) = \int_{\mathbf{M}^{m\times N}} \psi(A)\, d\nu_x(A).$$

The proof is a direct consequence of the previous theorem and is left to the reader.

We would also like to give some conditions that would improve weak to strong convergence. These involve the notion of strict quasiconvexity. We say that a continuous function defined over the space of matrices $\mathbf{M}^{m\times N}$ is uniformly strict $W^{1,p}$-quasiconvex if there is a $\gamma > 0$ such that

$$\varphi(Y) + \gamma \int_\Omega |\nabla u - Y|^p \, dx \leq \int_\Omega \varphi(\nabla u)\, dx,$$

for any $u \in W^{1,p}(\Omega)$ with $u - u_Y \in W_0^{1,p}(\Omega)$. This condition is connected to the study of partial regularity for minimizers of vector variational principles. A somewhat less restrictive notion, which we simply call strict $W^{1,p}$-quasiconvexity, requires that for any homogeneous $W^{1,p}$-parametrized measure, ν, with underlying deformation Y, the strict inequality

$$\varphi(Y) < \int_{\mathbf{M}^{m\times N}} \varphi(A)\, d\nu(A)$$

holds unless $\nu = \delta_Y$. The fact that we can understand the gradient of any $W^{1,p}(\Omega)$ function with affine boundary values Y as a homogeneous $W^{1,p}$-parametrized measure through the averaging procedure tells us that uniformly strict $W^{1,p}$-quasiconvexity truly implies strict $W^{1,p}$-quasiconvexity.

Theorem 3.16 *If $u_j \rightharpoonup u$ in $W^{1,p}(\Omega)$, φ is a continuous, strict $W^{1,p}$-quasiconvex function satisfying*

$$\max(0, c\,|A|^p - 1) \leq \varphi(A) \leq C(1 + |A|^p), \quad 0 < c \leq C,$$

and

$$\int_\Omega \varphi(\nabla u)\, dx = \lim_{j\to\infty} \int_\Omega \varphi(\nabla u_j)\, dx,$$

then $u_j \to u$ strongly in $W^{1,p}(\Omega)$.

Proof. Using the weak lower semicontinuity for any measurable $E \subset \Omega$ and its complement $\Omega \setminus E$ we obtain

$$\int_E \varphi(\nabla u)\, dx \leq \liminf_{j\to\infty} \int_E \varphi(\nabla u_j)\, dx,$$
$$\int_{\Omega\setminus E} \varphi(\nabla u)\, dx \leq \liminf_{j\to\infty} \int_{\Omega\setminus E} \varphi(\nabla u_j)\, dx.$$

Coupling these two inequalities, by hypothesis we get, in fact, equality for the whole sequence. Therefore we should have

$$\int_E \varphi(\nabla u)\, dx = \lim_{j\to\infty} \int_E \varphi(\nabla u_j)\, dx,$$

for any $E \subset \Omega$. On the other hand, by Theorem 3.5,

$$\varphi(\nabla u(x)) \leq \int_{\mathbf{M}^{m \times N}} \varphi(A)\, d\nu_x(A), \quad \text{a.e. } x \in \Omega,$$

if $\nu = \{\nu_x\}_{x \in \Omega}$ is the parametrized measure associated to $\{\nabla u_j\}$. For any E

$$\begin{aligned}\lim_{j \to \infty} \int_E \varphi(\nabla u_j)\, dx &= \int_E \varphi(\nabla u)\, dx \\ &\leq \int_E \int_{\mathbf{M}^{m \times N}} \varphi(A)\, d\nu_x(A)\, dx.\end{aligned}$$

But in fact, by Theorem 6.11, equality must hold

$$\lim_{j \to \infty} \int_E \varphi(\nabla u_j)\, dx = \int_E \int_{\mathbf{M}^{m \times N}} \varphi(A)\, d\nu_x(A)\, dx,$$

and in particular

$$\int_E \varphi(\nabla u)\, dx = \int_E \int_{\mathbf{M}^{m \times N}} \varphi(A)\, d\nu_x(A)\, dx.$$

Hence

$$\varphi(\nabla u(x)) = \int_{\mathbf{M}^{m \times N}} \varphi(A)\, d\nu_x(A), \quad \text{a.e. } x \in \Omega.$$

Since by the localization property, Theorem 8.4, each individual ν_x is a homogeneous $W^{1,p}$-parametrized measure, the strict $W^{1,p}$-quasiconvexity of φ, let us conclude that $\nu_x = \delta_{\nabla u(x)}$ for a.e. $x \in \Omega$. By Theorem 3.14, $\{\varphi(\nabla u_j)\}$ (or some appropriate subsequence) is weakly convergent in $L^1(\Omega)$. The coercivity assumed on φ enables us to obtain the equiintegrability on the powers $\{|\nabla u_j|^p\}$. Proposition 6.12 helps us in finishing the proof. ■

An immediate consequence of this theorem is that minimizing sequences for strict $W^{1,p}$-quasiconvex integrands with the proper bounds always converge strongly in $W^{1,p}(\Omega)$ to minimizers.

7. Bibliographical remarks

The weak lower semicontinuity property is bound to convexity properties. Almost all of the references related to quasiconvexity treat the problem of weak lower semicontinuity under several settings. [3], [4], [29], [33], [37], [38], [97], [106], [151], [166], [193], [201], [216], [220], [221], [225], [233], [238], [240], [269], [272], [289] is a partial list of some basic papers. The books referred to in Chapter 1 that take a point of view based on weak convergence are also references to keep in mind for convexity, quasiconvexity and lower semicontinuity. More general treatments of weak lower semicontinuity in the context of BV functions or with respect to measures have considered systematically by the Italian and the French schools: [12], [13], [14], [15], [16], [18], [52], [53], [54], [106], [141], to mention just a few. See also [143]. The approach taken here based on parametrized measures was mainly inspired by [186], [191], [260]. A similar point of view is taken in [40].

Polyconvexity was studied for the first time systematically in [29]. The fundamental observation that some of the difficulties in nonlinear elasticity are solved by considering polyconvex stored energy densities was also made in [29]. Existence theorems are usually obtained as indicated as soon as weak lower semicontinuity results are available. What is not so elementary is achieving existence results in spite of nonconvexity. For this problem there is abundant bibliography: [20], [21], [22], [23], [25], [26], [65], [66], [67], [68], [69], [99], [102], [117], [156], [217], [218], [223], [226], [227], [228], [229], [266], [267], [292], [300].

Quasiaffine functions have been also studied extensively because of their remarkable properties. Some works in this direction are: [33], [167], [168], [240], [241], [242], [269]-[272]. A remarkable weak lower semicontinuity counterexample for the jacobian is contained in [215].

Concerning continuum mechanics and elasticity, there are some good introductory textbooks like [78] and [163]. The material in Section 5 has been taken from [78]. Some other references in this direction are [17], [29], [36].

The results in the last section have been taken from [190]. The problem of the regularity of minimizers under conditions of strict convexity has been studied in [5], [30], [133], [134], [159], [222].

Chapter 4
Nonconvexity and Relaxation

1. Introduction

We have seen in the previous chapter that the weak lower semicontinuity property is crucial in order to employ the direct method of the calculus of variations to find minimizers of variational principles. This property is inherited by functionals whose integrands enjoy the appropriate convexity. Nonetheless, for an ever increasing number of interesting problems these convexity properties fail. In some cases, specific techniques may provide solutions to problems. In some others, this lack of convexity is a precursor of nonsolvability, at least in a classical sense. In the latter, highly oscillatory phenomena are usually involved. Parametrized measures were originally introduced by Young to account for oscillations in nonconvex optimal control problems where one could not reasonably expect classical solutions.

To explain the nature of oscillatory minimizing sequences, let us look at a very simple, one-dimensional example essentially due to Bolza. The functional is

$$I(u) = \int_0^1 \left[\varphi(u'(x)) + u(x)^2\right] \, dx,$$

where $\varphi(t) = (t^2 - 1)^2$. The boundary conditions we take for admissibility are $u(0) = u(1) = 0$. We notice that there is some kind of conflict between the two contributions to I. On the one hand the term on the derivative would favor slopes $+1$ or -1, while the quadratic term would be minimized when $u = 0$ which does not have either of the two slopes. This incompatibility is responsible for the behavior of minimizing sequences and ultimately for the lack of minimizers for I. This idea can be made precise in the following way.

Step 1. The infimum of I under the given boundary conditions is 0. For this, it is enough to take the saw-tooth functions

$$u_k(x) = \frac{1}{2k} - \left| x - \frac{\langle kx \rangle}{k} - \frac{1}{2k} \right|.$$

(Remember that $\langle a \rangle$ stands for the integer part of $a \in \mathbf{R}$). It is an easy exercise to check that this formula gives functions with slopes $+1$ and -1 and fulfills the boundary conditions. Moreover, since $\{u_k\}$ converges strongly in the L^∞-norm to 0 we get $I(u_k) \to 0$. Indeed $\{u_k\}$ converges weakly in $W^{1,\infty}(0,1)$ to 0.

Step 2. I does not admit minimizers because if there exists a function u such that $I(u) = 0$, on the one hand $u \equiv 0$ but on the other hand the slopes that u may use should be restricted to ± 1 and this is obviously impossible. In particular, even though the minimizing sequence $\{u_k\}$ converges weakly to 0, the function 0 is not a minimizer so that I cannot be weakly lower semicontinuous. Observe that φ is not convex. The behavior of any minimizing sequence for this problem is essentially the one described by $\{u_k\}$.

The analysis of I may now proceed in two directions, although they are not, in fact, different directions. First, we can "convexify" the integrand φ and consider the regular variational problem associated with the convexification of φ,

$$\overline{I}(u) = \int_0^1 \left[\varphi^{**}(u'(x)) + u(x)^2\right] dx,$$

where φ^{**} is the convexification of φ. This new problem is usually referred to as the associated relaxed problem. The connection between the relaxed problem and the original one is established through the fact that the two infima, the one for I and the one for $\overline{I}$, are equal (under suitable technical assumptions). In this specific case

$$\varphi^{**}(t) = \begin{cases} \varphi(t), & |t| \geq 1, \\ 0, & |t| \leq 1, \end{cases}$$

so that the two infima are easily seen to vanish. Notice that the function that identically vanishes in $[0,1]$ is a minimizer for $\overline{I}$, because $\overline{I}$ is weak lower semicontinuous, but it is not for I.

The relaxed problem does not provide much information on the behavior of minimizing sequences. If one desires insight on that matter, the original problem must be embedded in a generalized variational principle where parametrized measures are allowed to compete in the minimization process. The functional becomes

$$\tilde{I}(\nu) = \int_0^1 \left[\int_{\mathbf{R}} \varphi(\lambda)\, d\nu_x(\lambda) + u(x)^2\right] dx,$$

$$\nu = \{\nu_x\}_{x\in\Omega}, \quad \int_0^1 \int_{\mathbf{R}} \lambda\, d\nu_y(\lambda)\, dy = 0,$$

where

$$u(x) = \int_0^x \int_{\mathbf{R}} \lambda\, d\nu_y(\lambda)\, dy.$$

Notice that

$$\tilde{I}(\delta_{u'(x)}) = I(u),$$

for any admissible u. In this one-dimensional example there are no more restrictions on the competing ν. In the vector case and in more general situations, understanding these constraints is the most important and most delicate question concerning the analysis of $\tilde{I}$. Under suitable hypotheses, $\tilde{I}$ admits always a parametrized measure solution and all admissible generating sequences are minimizing sequences for the original problem. In this sense we say that the behavior of minimizing sequences is encoded in the parametrized measure solution for $\tilde{I}$.

The oscillatory behavior we are attempting to examine is present in many interesting situations in models of continuum mechanics, problems of optimal design, optimal control and even turbulence phenomena and large time behavior for complicated systems. Some of these have already been examined in Chapter 2. They have served to motivate our way of understanding weak convergence through parametrized measures. In the next chapter some other fundamental situations are analyzed in greater detail.

This chapter is devoted to the analysis of nonconvexity for the standard problem of the calculus of variations. A basic relaxation theorem is proved in Section 2 based on the program set forth in Chapter 1. An analysis of variational principles where we let parametrized measures compete follows. The main underlying ideas might also be applied to other problems that may not exactly fit in this category, as has been pointed out in the previous paragraph.

2. Relaxation theorem

A typical relaxation theorem establishes, under some technical assumptions, that the infimum of any functional does not change when we replace the integrand by its "convexification".

Theorem 4.1 *Let $\varphi : \Omega \times \mathbf{R}^m \times \mathbf{M}^{m\times N} \to \mathbf{R}$ be a Carathéodory function, such that*

$$c(|A|^p - 1) \le \varphi(x, \lambda, A) \le h(x, u)(1 + |\lambda|^p + |A|^p),$$

where $0 < c$, $p > 1$ and h is a locally bounded function. For any given $u_0 \in W^{1,p}(\Omega)$, the two infima

$$\inf\left\{\int_\Omega \varphi(x, u(x), \nabla u(x))\,dx : u - u_0 \in W_0^{1,p}(\Omega)\right\},$$

$$\inf\left\{\int_\Omega Q\varphi(x, u(x), \nabla u(x))\,dx : u - u_0 \in W_0^{1,p}(\Omega)\right\},$$

are equal.

The basic fact we need to prove this relaxation theorem is contained in the following lemma which by itself is a homogeneous version of the theorem.

Lemma 4.2 *Let $\psi : \mathbf{M}^{m\times N} \to \mathbf{R}$ be continuous such that*

$$c(|A|^p - 1) \le \psi(A) \le C(1 + |A|^p), \quad p > 1, 0 < c < C.$$

For any $Y \in \mathbf{M}^{m\times N}$ there exists a homogeneous $W^{1,p}$-parametrized measure, ν_Y, such that

$$Y = \int_{\mathbf{M}^{m\times N}} A\, d\nu_Y(A), \quad Q\psi(Y) = \int_{\mathbf{M}^{m\times N}} \psi(A)\, d\nu_Y(A).$$

Moreover

$$\int_{\mathbf{M}^{m\times N}} |A|^p\, d\nu_Y(A) < \infty.$$

Proof of lemma. Consider the following variational principle

$$Q\psi(Y) = \inf\left\{\frac{1}{|\Omega|}\int_\Omega \psi(\nabla u)\, dx : u \in W^{1,p}(\Omega), u - u_Y \in W_0^{1,p}(\Omega)\right\},$$

and let $\{u_j\}$ be a minimizing sequence. Since we have affine boundary conditions, by the average process Theorem 8.1, we may assume that the $W^{1,p}$-parametrized measure associated to $\{\nabla u_j\}$ is homogeneous, ν_Y, so that $u_j \rightharpoonup u_Y$ in $W^{1,p}(\Omega)$. Then

$$Y = \int_{\mathbf{M}^{m\times N}} A\, d\nu_Y(A).$$

By Lemma 8.10 and Theorem 8.13, $Q\psi$ is quasiconvex, so that

$$Q\psi(Y) = \inf\left\{\frac{1}{|\Omega|}\int_\Omega Q\psi(\nabla u)\, dx : u \in W^{1,p}(\Omega), u - u_Y \in W_0^{1,p}(\Omega)\right\}. \quad (4\text{–}1)$$

Since for all j

$$Q\psi(Y) \le \frac{1}{|\Omega|}\int_\Omega Q\psi(\nabla u_j)\, dx \le \frac{1}{|\Omega|}\int_\Omega \psi(\nabla u_j)\, dx,$$

we conclude that $\{u_j\}$ is also minimizing in (4–1). By Lemma 8.12 $Q\psi$ inherits the same coercivity than ψ because the lower bound for ψ is a convex function and hence we have exactly the same lower bound for $Q\psi$. Theorem 3.14 enables us to affirm that

$$Q\psi(\nabla u_j) \rightharpoonup Q\psi(Y) \quad \text{in } L^1(\Omega).$$

By the coercivity $\{|\nabla u_j|^p\}$ (or some subsequence) also converges weakly in $L^1(\Omega)$. By the upper bound on ψ, the same is true for $\{\psi(\nabla u_j)\}$ and hence we have the representation

$$|\Omega|\, Q\psi(Y) = \lim_{j\to\infty}\int_\Omega \psi(\nabla u_j)\, dx = |\Omega| \int_{\mathbf{M}^{m\times N}} \psi(A)\, d\nu_Y(A),$$

as desired. The fact that $|\cdot|^p$ is integrable with respect to ν_Y is an immediate consequence of the bounds assumed on ψ. ∎

Proof of Theorem 4.1. Let m and Qm denote the two infima, respectively. Trivially $Qm \leq m$. In order to show equality, let u be any admissible function in $W^{1,p}(\Omega)$ so that $u - u_0 \in W_0^{1,p}(\Omega)$. By the bounds assumed on φ and Lemma 4.2 we can find for a.e. $x \in \Omega$, a homogeneous $W^{1,p}$-parametrized measure, ν^x, such that

$$\nabla u(x) = \int_{\mathbf{M}^{m\times N}} A \, d\nu^x(A),$$

$$Q\varphi(x, u(x), \nabla u(x)) = \int_{\mathbf{M}^{m\times N}} \varphi(x, u(x), A) \, d\nu^x(A).$$

Consider the family of probability measures $\nu = \{\nu^x\}_{x\in\Omega}$. We would like to show that ν is a $W^{1,p}$-parametrized measure. According to Theorem 8.16 we have to check three conditions. These hold essentially by construction. First, the fact that Jensen's inequality holds for quasiconvex functions in $\mathcal{E}^p$ is true because each ν^x has been chosen to be a homogeneous $W^{1,p}$-parametrized measure. The compatibility condition that the first moment should be a gradient is also automatic. Finally the coercivity condition assumed on φ yields the finiteness of the integral of the pth power against ν. Thus, there exists a sequence of functions in $W^{1,p}(\Omega)$, $\{u_j\}$, whose parametrized measure is precisely $\nu = \{\nu^x\}_{x\in\Omega}$ and $\{|\nabla u_j|^p\}$ is weakly convergent in $L^1(\Omega)$ (Lemma 8.15). Once we have this weak convergence, we can assume that each u_j is admissible by Lemma 8.3. In this case

$$\lim_{j\to\infty} \int_\Omega \varphi(x, u_j(x), \nabla u_j(x)) \, dx = \int_\Omega \int_{\mathbf{M}^{m\times N}} \varphi(x, u(x), A) \, d\nu^x(A) \, dx$$
$$= \int_\Omega Q\varphi(x, u(x), \nabla u(x)) \, dx.$$

The arbitrariness of u yields the result. ∎

3. Parametrized measures solutions of variational principles

We have already talked about parametrized measures solutions of variational principles in some of the examples in Chapter 2. We would like to examine from this point of view the standard problem of the calculus of variations under failure of the quasiconvexity condition for the integrand. Important applications will be discussed in Chapter 5.

In many different models of mathematical physics we need to consider variational principles where the integrand φ of the energy functional

$$I(u) = \int_\Omega \varphi(x, u(x), \nabla u(x)) \, dx, \qquad u \in W^{1,p}(\Omega), u - u_0 \in W_0^{1,p}(\Omega),$$

is not quasiconvex on the gradient variable. u_0 is assumed to be some fixed function in $W^{1,p}(\Omega)$. As pointed out, the typical behavior of minimizing sequences for these functionals is highly oscillatory: while the oscillations take place in regions of increasing fineness they remain of finite, nonvanishing amplitude. In these circumstances we talk about parametrized measures solutions. The assumptions for φ are the usual bounds

$$c(|A|^p - 1) \leq \varphi(x, \lambda, A) \leq C(1 + |\lambda|^p + |A|^p), \quad 0 < c \leq C.$$

We would like to allow $W^{1,p}$-parametrized measures to compete in the energy minimization process. In order to do this, we define the energy of such a parametrized measure by

$$\tilde{I}(\mu) = \int_\Omega \int_{\mathbf{M}^{m\times N}} \varphi(x, u(x), A)\, d\mu_x(A)\, dx,$$

where $\mu = \{\mu_x\}_{x\in\Omega}$ is a $W^{1,p}$-parametrized measure generated by a sequence of gradients in $W^{1,p}(\Omega)$, subject to the compatibility conditions

$$\nabla u(x) = \int_{\mathbf{M}^{m\times N}} A\, d\mu_x(A), \qquad u - u_0 \in W_0^{1,p}(\Omega),$$

giving the relationship between u and μ. We say that such a μ is admissible for $\tilde{I}$.

Note that we can always take $\mu_x = \delta_{\nabla u(x)}$ for some admissible u and in this case $\tilde{I}(\mu) = I(u)$.

Lemma 4.3 *I admits a minimizing sequence $\{u_k\}$ such that $\{|\nabla u_k|^p\}$ is weakly convergent in $L^1(\Omega)$.*

Proof. Let $\{v_k\}$ be any minimizing sequence for I. By the bounds assumed on φ, it is a bounded sequence in $W^{1,p}(\Omega)$. Let $\nu = \{\nu_x\}_{x\in\Omega}$ denote the $W^{1,p}$-parametrized measure associated to the sequence of gradients $\{\nabla v_k\}$. By Lemma 8.15, ν can also be generated by some other sequence of gradients $\{\nabla w_k\}$ such that $\{|\nabla w_k|^p\}$ is weakly convergent in $L^1(\Omega)$. In particular, both sequences have the same weak limit in $W^{1,p}(\Omega)$, u, and $w_k \to u$ strong in $L^p(\Omega)$. By Lemma 8.3 we can find $\{u_k\}$ admissible for I and still have the equiintegrability of $\{|\nabla u_k|^p\}$.

Since $\{v_k\}$ is minimizing

$$\begin{aligned}\lim_{k\to\infty} \int_\Omega \varphi(x, v_k(x), \nabla v_k(x))\, dx &\leq \lim_{k\to\infty} \int_\Omega \varphi(x, u_k(x), \nabla u_k(x))\, dx \\ &= \int_\Omega \int_{\mathbf{M}^{m\times N}} \varphi(x, u(x), A)\, d\nu_x(A)\, dx.\end{aligned}$$

By Theorem 6.11 strict inequality in the first two terms is impossible, so that $\{u_k\}$ is also minimizing. ∎

With this lemma we can now prove the following theorem.

Theorem 4.4

$$\inf I(u) = \inf \tilde{I}(\mu) = \inf \overline{I}(u),$$

where $\overline{I}(u)$ is the energy functional whose energy density is the quasiconvexification of φ with respect to the gradient variable.

Proof. Let m, $\tilde{m}$ and $\overline{m}$ denote the three infima, respectively. In the previous section we have already shown that $m = \overline{m}$.

By the observation made prior to Lemma 4.3 we conclude that $\tilde{m} \leq m$. To show equality, let μ be admissible for $\tilde{I}$. By Lemmas 8.15 and 8.3, we can find a sequence of $W^{1,p}(\Omega)$-functions, $\{u_j\}$, such that $u_j - u_0 \in W_0^{1,p}(\Omega)$, $\{|\nabla u_j|^p\}$ is weakly convergent in $L^1(\Omega)$ and the parametrized measure associated to $\{\nabla u_j\}$ is μ. Thus

$$\begin{aligned}\lim_{j\to\infty} I(u_j) &= \lim_{j\to\infty} \int_\Omega \varphi(x, u_j(x), \nabla u_j(x))\, dx \\ &= \int_\Omega \int_{\mathbf{M}^{m\times N}} \varphi(x, u(x), A)\, d\mu_x(A)\, dx \\ &= \tilde{I}(\mu),\end{aligned}$$

where

$$\nabla u(x) = \int_{\mathbf{M}^{m\times N}} A\, d\mu_x(A).$$

This clearly implies that $m = \tilde{m}$. ∎

The advantage of dealing with $\tilde{I}$ is that it admits minimizers within the class of $W^{1,p}$-parametrized measures only under the usual bounds on φ. No convexity condition is needed or assumed.

Corollary 4.5 *There exists a ν admissible for $\tilde{I}$ such that*

$$\tilde{I}(\nu) = \inf \tilde{I}(\mu).$$

Proof. Take a minimizing sequence for I, $\{u_j\}$, and let ν be the parametrized measure associated to $\{\nabla u_j\}$. By Lemma 4.3 we may assume without loss of generality that $\{|\nabla u_j|^p\}$ is weakly convergent in $L^1(\Omega)$, so that

$$\tilde{m} = m = \lim_{j\to\infty} I(u_j) = \tilde{I}(\nu).$$ ∎

In this way we have a limit energy density for I

$$\overline{\varphi}(x) = \int_{\mathbf{M}^{m\times N}} \varphi(x, u(x), A)\, d\nu_x(A),$$

where $\nu = \{\nu_x\}_{x\in\Omega}$ is a minimizer for $\tilde{I}$. Moreover for any continuous, nonlinear quantity $\psi : \mathbf{M}^{m\times N} \to \mathbf{R}$ such that

$$|\psi(A)| \le C(1 + |A|^p),$$

we have a representation in terms of ν

$$\lim_{j\to\infty} \int_\Omega \psi(\nabla u_j)\, dx = \int_\Omega \overline{\psi}(x)\, dx = \int_\Omega \int_{\mathbf{M}^{m\times N}} \psi(A)\, d\nu_x(A)\, dx.$$

Given a non-convex functional I, we now have two ways to obtain a well-behaved functional associated with it, $\overline{I}$ and $\tilde{I}$. A natural question is how minimizers for both functionals are related.

Corollary 4.6 *Let ν be a minimizer for $\tilde{I}$. If*

$$\nabla u(x) = \int_{\mathbf{M}^{m\times N}} A\, d\nu_x(A), \qquad \text{a.e. } x \in \Omega, \tag{4–2}$$

for $u \in W^{1,p}(\Omega)$, then u is a minimizer for $\overline{I}$ and

$$Q\varphi(x, u(x), \nabla u(x)) = \int_{\mathbf{M}^{m\times N}} \varphi(x, u(x), A)\, d\nu_x(A), \qquad \text{a.e. } x \in \Omega. \tag{4–3}$$

Conversely, if u is a minimizer for $\overline{I}$ and $\nu = \{\nu_x\}_{x\in\Omega}$ is an admissible $W^{1,p}$-parametrized measure such that (4–2) and (4–3) hold, then ν is a minimizer for $\tilde{I}$.

The proof is elementary. Simply notice that u is admissible, and by Jensen's inequality we can write the following chain of inequalities

$$\begin{aligned}
\overline{m} \le \overline{I}(u) &= \int_\Omega Q\varphi(x, u(x), \nabla u(x))\, dx \\
&\le \int_\Omega \int_{\mathbf{M}^{m\times N}} Q\varphi(x, u(x), A)\, d\nu_x(A)\, dx \\
&\le \int_\Omega \int_{\mathbf{M}^{m\times N}} \varphi(x, u(x), A)\, d\nu_x(A)\, dx \\
&= \tilde{I}(\nu) = \tilde{m} = \overline{m}.
\end{aligned}$$

Therefore u is a minimizer for $\overline{I}$ and (4–3) must hold true. The same is true for the converse.

Finally, we give some information about the support of the parametrized measure minimizer.

Corollary 4.7

$$\textit{supp}\,(\nu_x) \subset \{\varphi(x, u(x), \cdot) = Q\varphi(x, u(x), \cdot)\}\,, \qquad \text{a.e. } x \in \Omega.$$

Proof. Observe that by the relaxation Theorem 4.1,

$$\int_\Omega \int_{\mathbf{M}^{m\times N}} [\varphi(x,u(x),A) - Q\varphi(x,u(x),A)]\, d\nu_x(A)\, dx = 0,$$

and the integrand is nonnegative. Therefore the support of ν_x should be contained where the integrand vanishes. ■

Let us once again emphasize the importance of understanding the restrictions that govern parametrized measures that may compete in the variational principle for $\tilde{I}$. If one forgets this issue, the connection between both variational principles, the one for I and the one for $\tilde{I}$ may be lost, and information for I may not be recovered from $\tilde{I}$ if the analysis overlooks those restrictions. As a matter of fact, this is the heart of the problem of understanding relaxation and was one of our main motivations in investigating characterizations for parametrized measures generated by gradients.

Let us close this chapter by looking at the one-dimensional example mentioned in the introduction. The variational principle is

$$I(u) = \int_0^1 \left[\varphi(u'(x)) + (u(x) - f(x))^2\right] dx,$$
$$u \in H^1(0,1), \quad u(0) = u_0, u(1) = u_1,$$

where $\varphi(\lambda) = (|\lambda| - 1)^2$ is the usual nonconvex, double well potential and $f : [0,1] \to \mathbf{R}$ is some specific, smooth, bounded function. All the necessary hypotheses hold for $p = 2$. The associated functionals $\tilde{I}$ and $\overline{I}$ are given by

$$\tilde{I}(\nu) = \int_0^1 \left[\int_{\mathbf{R}} \varphi(\lambda)\, d\nu_x(\lambda) + (u(x) - f(x))^2\right] dx,$$
$$\nu = \{\nu_x\}_{x\in\Omega}, \quad \int_0^1 \int_{\mathbf{R}} \lambda\, d\nu_y(\lambda)\, dy = u_1 - u_0,$$

where

$$u(x) = u_0 + \int_0^x \int_{\mathbf{R}} \lambda\, d\nu_y(\lambda)\, dy.$$

In the one-dimensional case the assumptions on the admissible μ are less restrictive since the condition that the first moment of μ be a gradient is always true. For $\overline{I}$ we get

$$\overline{I}(u) = \int_0^1 \left[\varphi^{**}(u'(x)) + (u(x) - f(x))^2\right] dx,$$
$$u \in H^1(0,1), \quad u(0) = u_0, u(1) = u_1.$$

Observe that the second term in the integral for $\overline{I}$ is strictly convex, thus making the minimizer for $\overline{I}$ unique. Let u denote such minimizer. According to

Corollary 4.6, minimizers for $\tilde{I}$ are obtained by seeking the family of probability measures ν such that

$$\varphi^{**}(u'(x)) = \int_{\mathbf{R}} \varphi(\lambda)\, d\nu_x(\lambda), \quad \text{a.e. } x \in (0,1).$$

In this simplified situation it is easy to observe that given any real number $u'(x)$ there is a unique ν_x verifying the previous condition. Indeed we can write

$$\nu_x = \begin{cases} \lambda(x)\delta_1 + (1-\lambda(x))\delta_{-1}, & |u'(x)| \le 1, \\ \delta_{u'(x)}, & |u'(x)| \ge 1, \end{cases}$$

$$\lambda(x) = \frac{1+u'(x)}{2}.$$

This family of probability measures is the unique minimizer for $\tilde{I}$.

4. Bibliographical remarks

Variational problems that lack convexity have attracted researchers over the years. This is all the more so because of the interesting applications that such analysis for nonconvex problems has. Relaxation theorems and convex envelopes in several frameworks are very well understood by now. The literature on this topic is copious. We do not claim to include all the relevant papers here. Some of them deal with different situations and need delicate techniques, especially those related to BV functions and measures. See [2], [3], [15], [54], [62], [91], [92], [94], [98], [104], [105], [137], [138], [140], [144], [152], [194], [199], [224], [280], [293], [306], [312]. See also [197]. Many of the textbooks mentioned in Chapter 1 include some treatment of nonconvexity and relaxation.

The generalized variational principle in terms of parametrized measures goes back to [314] and [315], and was described and analyzed in some detail in [77] in a framework similar to ours. Related works include [148], [179], [190], [243].

The numerical analysis of nonconvex problems has received much attention lately. Nevertheless, we lack efficient algorithms to compute oscillations. Because these take place in so small a scale, computers have a lot of trouble detecting them in an accurate way. References dealing with this topic are [57], [63], [74], [85], [147], [211], [251], [252], [261], [263], [281].

Chapter 5
Phase Transitions and Microstructure

1. Two main examples from continuum mechanics

We have tried to emphasize in the previous chapter the importance of the study of variational principles for which some lack of convexity leads one to consider the behavior of minimizing sequences. From the mathematical point of view, there are two ways to proceed whenever there are no minimizers as a consequence of this lack of convexity. One is to "convexify" the energy density itself or the nonconvex constraints involved in order to obtain a new functional which can be analyzed through the techniques dicussed in Chapter 3. The task is to relate the information concerning this convexified functional with the original one. Relaxation theorems refer to this issue. Another possibility is to enlarge sufficiently the class of competing objects in some kind of generalized variational setting as to include minimizers. These generalized objects are parametrized measures. They were introduced by Young in this same context to understand ill-posed variational problems.

The type of oscillatory phenomena described by means of the one-dimensional example in Chapter 4 is also present in martensitic transformations where the oscillations of the deformation gradient, in the context of nonlinear elasticity, remain finite in amplitude but take place in smaller and smaller spatial scales. This extremely fine structure of alternate layers has been referred to as *microstructure*, a term that accepts many different meanings but intuitively reflects the behavior of minimizing sequences. These models are placed in the framework of nonlinear elasticity and the connection between continuum models and crystallographic properties of materials is made through the Cauchy-Born rule that postulates the existence of a continuous, nonnegative energy density, φ, that provides a measure of energy corresponding to a deformed crystal lattice. The basic axiom of elasticity theory (see Section 5, Chapter 3) is that the total free energy can be represented as the integral over the reference configuration Ω of the local density associated with a deformation of the body $u : \Omega \to \mathbf{R}^3$,

$$I(u) = \int_\Omega \varphi(\nabla u)\, dx.$$

∇u is the deformation gradient and represents a measure of the local strain around each point $x \in \Omega$ (for the purpose of this discussion, temperature is assumed to be held constant). From the physical point of view, φ should incorporate frame indifference and reflect material symmetry as well. These facts get translated into the invariance

$$\varphi(QFH) = \varphi(F), \quad Q \in SO(3), H \in \mathcal{P}, \tag{5–1}$$

that arise from the Cauchy-Born rule, where $\mathcal{P}$ is a set of matrices reflecting the crystalline symmetry of the material (in situations of interest $\mathcal{P}$ is a finite group of matrices reflecting the symmetries of one of the phases taken as reference). The consequences of (5–1) are crucial, namely, that invariance is responsible for lack of quasiconvexity for φ and ultimately for the presence of microstructure in this type of problem. Indeed, suppose that for a particular matrix F with zero energy, $\varphi(F) = 0$, we can find $Q \in SO(3)$ and $H \in \mathcal{P}$ such that F and QFH are rank-one related

$$F - QFH = a \otimes n. \tag{5–2}$$

By the invariance (5–1), $\varphi(QFH) = 0$ and were φ rank-one convex, we would have (bearing in mind that $\varphi \geq 0$) that φ vanishes along the segment joining F and QFH. This means that any convex combination of F and QFH has zero energy and hence it should be contained in the zero set of φ. In the situation we are discussing on martensitic transformations this is not so and hence φ cannot be quasiconvex. The zero set of φ plays a fundamental role since we can look for minimizers or minimizing sequences whose gradients take on values in this set as often as possible. If (5–2) holds, by the basic construction described in Chapter 1 related to rank-one convexity, for any $t \in (0,1)$ we can find a sequence of Lipschitz deformations, $\{u_j\}$, such that $u_j - u_Y \in W_0^{1,\infty}(\Omega)$ where $u_Y(x) = Yx$ is affine, $Y = tF + (1-t)QFH$ and ∇u_j takes on the values F and QFH in alternate layers with normal n and relative frecuency t and $1-t$, respectively. This sequence of deformations is minimizing, $I(u_j) \to 0$, and represents a *stress-free microstructure.* (5–2) is the basic equation of the crystallographic theory of martensite and it can be derived rigorously from energy considerations.

The invariance (5–1) gives a lot of information about the structure of the zero set of φ. The typical situation is the following. The set $\mathcal{P}$ is a discrete group of several matrices accounting for the symmetry of the material we are working with. Assume that we take a particular affine, homogeneous deformation with minimum energy as a reference so that $\varphi(\mathbf{1}) = 0$ where $\mathbf{1}$ is the identity matrix. For each $\mathbf{1}H$, $H \in \mathcal{P}$, we have a potential well $\{Q\mathbf{1}H : Q \in SO(3)\}$ made up of minimum energy matrices, each one a copy of $SO(3)$. Altogether we obtain a finite number of potential wells which contain no segment. Under these circumstances minimizing sequences for the internal energy functional will develop oscillations taking place in a very fine scale as announced.

A different source of nonconvexity may be located on the set of competing functions in a particular variational principle, so that even if the functional itself is convex, the analysis leads one to consider some kind of relaxed formulation. One such interesting example that we will analyze is some detail through divergence-free parametrized measures comes from the theory of micromagnetics. Micromagnetics is a mathematical model of ferromagnetism intended to provide a description of the magnetization of a ferromagnetic body under the action of an external applied field. The theory has evolved to seek an explanation for the fine structures observed in experiments. We will restrict our attention to the rigid case in which the only state variable is the magnetization m, assumed to be a vector field over the body. One interesting assumption is that the magnetization field is assumed to be of constant length if we do not allow temperature variations. For simplicity we set $|m| = 1$. This hypothesis reflects the local saturation of the material. The variational principle governing equilibrium configurations of a large body for the magnetization m consists of several terms which give rise to the following energy functional

$$I(m) = \int_\Omega \varphi(m)\, dx + \frac{1}{2}\int_{\mathbf{R}^N} |\nabla u|^2\, dx - \int_\Omega H \cdot m\, dx, \tag{5–3}$$

subject to the constraints

$$|m| = 1, \quad \text{a.e. } x \in \Omega,$$
$$\operatorname{div}(-\nabla u + m\chi_\Omega) = 0 \quad \text{in } H^{-1}(\mathbf{R}^N).$$

Notice that the solution operator to the differential constraint, which can be solved by means of the Lax-Milgram theorem, is weak continuous from $L^2(\Omega)$ to $H^1(\Omega)$. The three terms in the energy functional correspond respectively to the anisotropy energy accounting for the tendency of the material to align the magnetization with specific crystallographic directions, the magnetostatic energy and the interaction energy due to the presence of an external, constant applied field H. Two cases are of particular relevance concerning the anisotropy energy density φ: the uniaxial case when φ is minimized only along one particular direction and the cubic case where φ is minimized along three linearly independent directions. Let us write $\mathcal{A}$ for the set of magnetizations verifying these constraints. The aim of micromagnetics is to look for a magnetic field m which solves the problem

$$I(m) = \inf_{\mathcal{A}} I(u). \tag{5–4}$$

Such a field m, were it to exist, is said to "attain" the minimum energy. We will be mainly considering conditions under which non-attainment of minima occurs and provide some insight on how the behavior of such materials can be analyzed.

Suppose we were to use the direct method to look for minimizers of (5–3) and (5–4). We would find a minimizing sequence, $\{m^k\}$. Because of the admissibility condition $|m^k(x)| = 1$ a.e. $x \in \Omega$, we could choose a subsequence, not relabeled, converging weakly to some $m \in L^\infty(\Omega)$. This m would not be admissible unless $\{m^k\}$ converges strongly in $L^\infty(\Omega)$ to m. Said differently, if we insist on having $|m(x)| = 1$ a.e. $x \in \Omega$ then $\{m^k\}$ must converge strongly. Roughly speaking, this is due to the fact that the unit sphere contains no segment. There is no a priori reason why to expect that minimizing sequences converge strongly, and in fact they do not. The direct method is useless in this framework. This does not imply, however, that the variational principle (5–3) does not admit minimizers. In some cases the minimum energy is attained. On the other hand, there are several places where one can find explicit examples in which minimizing sequences develop finer and finer oscillations. They certainly converge only in the weak sense so that weak limits cannot represent physical magnetizations under the saturation constraint $|m| = 1$, as indicated. This nonconvex constraint pushes us to consider some kind of relaxed formulation in terms of parametrized measures associated to sequences of magnetizations. Notice that the convexity or nonconvexity of φ does not play a role here.

2. Phase transitions and microstructure

One of the most interesting and fascinating examples of nonconvexity in nonlinear elasticity is the case of solid-solid phase transformations in ordered materials. The behavior of such materials can be understood and analyzed as an energy minimization process. The main feature of these materials is an extremely fine mixture of the phases. This property is a consequence of the lack of "convexity" of the energy density, and this in turn comes from symmetry considerations. In order to understand the mathematical analysis behind this phenomenon, we do not need to know the specific data for the materials under analysis but rather the key facts responsible for its qualitative behavior.

Assume we are dealing with a crystal considered as an elastic material. We postulate the existence of a nonnegative, free energy φ that depends on the change of shape (the deformation gradient) and on temperature as well. This is, more or less, the main consequence of the Cauchy-Born rule mentioned earlier. Above a certain critical temperature θ_0, there is a stable phase, taken as reference. By stable we simply mean that it minimizes φ, so that $\varphi_\theta(\mathbf{1}) = 0$, where $\mathbf{1}$ is the identity matrix and $\theta > \theta_0$. At the transition temperature θ_0 there is a change of stability or of crystal structure, so that below θ_0 the stable phase is not represented by $\mathbf{1}$ any more but by some other matrix U_0 describing the change in crystal structure that has taken place. Thus $\varphi_\theta(U_0) = 0$ but $\varphi_\theta(\mathbf{1}) > 0$ for $\theta < \theta_0$. At the transition temperature $\theta = \theta_0$ both phases may coexist $\varphi_{\theta_0}(U_0) = \varphi_{\theta_0}(\mathbf{1}) = 0$. Since the energy density $\varphi = \varphi_{\theta_0}$ must be frame indiferent

$$\varphi(RF) = \varphi(F)$$

for any rotation R. This implies that $\varphi(R) = \varphi(RU_0) = 0$ for all rotations R. We also have to take into account that the phase represented by $\mathbf{1}$ may have some symmetry. In fact, there is a group of rotations $\mathcal{P}$ of finite order representing the symmetry of the phase taken as reference. For example, if the atoms for the reference phase aligned themselves on cubic cells, then $\mathcal{P}$ would consists of the 24 rotations that leave a cube invariant. We should have at the critical temperature

$$\varphi(RU_0R^T) = \varphi(U_0) = 0,$$

for any $R \in \mathcal{P}$. Altogether we have found many matrices for which the free energy density φ vanishes:

$$\mathbf{1}, RR_1U_0R_1^T, RR_2U_0R_2^T, \ldots, RR_nU_0R_n^T,$$

where $\mathcal{P} = \{\mathbf{1}, R_1, R_2, \ldots, R_n\}$ and R is any rotation. We call each one of the sets $\{RR_iU_0R_i^T : R \text{ any rotation}\}$ a potential well associated to $U_i = R_iU_0R_i^T$ and make the further assumption that the free energy density φ is positive outside the set of the wells: the zero set for φ, $\{\varphi = 0\}$, is exactly the set of the wells.

Under these circumstances, we are looking for minimizers of the energy functional

$$I(u) = \int_\Omega \varphi(\nabla u)\,dx,$$

among all deformations u of the reference configuration $\Omega \subset \mathbf{R}^3$ satisfying appropriate boundary conditions. We have already explained why symmetry considerations lead to conclude that φ cannot be quasiconvex, and hence existence of minimizers is not guaranteed, at least it cannot be achieved by the direct method. In fact, fine phase mixtures provide minimizing sequences for which the weak limit is not a minimizer.

We can consider the setting described in the previous chapter where we would allow gradient parametrized measures to compete in the minimization process. Assume that some $u_0 \in W^{1,p}(\Omega)$ determines the prescribed boundary values we are interested in. Our generalized variational principle would read

$$\tilde{I}(\nu) = \int_\Omega \int_{\mathbf{M}^{3\times 3}} \varphi(A)\,d\nu_x(A)\,dx$$

$$\nabla u(x) = \int_{\mathbf{M}^{3\times 3}} A\,d\nu_x(A), \quad u - u_0 \in W_0^{1,p}(\Omega),$$

where $\nu = \{\nu_x\}_{x\in\Omega}$ is a $W^{1,p}$-parametrized measure and we assume p to be large enough. Under suitable bounds for φ there are solutions to this variational problem.

The structure of the zero set for φ plays a fundamental role if we are interested in finding equilibrium configurations, including parametrized measures minimizers, with zero energy. It is trivial to realize that this amounts to finding gradient parametrized measures whose support is contained in the set of wells for φ. It is of paramount importance to pay close attention to the fact that admissible parametrized measures for $\tilde{I}$ must be generated by gradients. Otherwise the variational principle would not have any significance with respect to the physical situation that we propose to examine. In obtaining useful information about the fact that parametrized measures should come from gradients, the so-called minor relations have been successfully exploited in many situations. These are nothing more than Jensen's inequality for the quasiaffine functions: the minors. Because these minors are continuous with respect to weak convergence in Sobolev spaces, Jensen's inequality is in fact an equality

$$M(\nabla u(x)) = \int_{\mathbf{M}^{3\times 3}} M(A)\, d\nu_x(A).$$

This has to hold for every minor M and a.e. $x \in \Omega$ if $\nu = \{\nu_x\}_{x\in\Omega}$ is admissible for $\tilde{I}$. These relations yield important information in the form of necessary conditions that in some cases are sufficient to find gradient parametrized measures supported on the set of wells for φ. For instance, by using the minor relations one can conclude that the existence of nontrivial microstructures is only possible if the wells are compatible. In the case of two wells this means that two matrices, one of each well, should differ by a rank-one matrix. If the energy density only has one well then there cannot be non-trivial microstructures.

The structure of the set of such gradient parametrized measures is extremely complex even for just two wells, and the problem is still far from being completely solved. Most of the microstructures observed in experiments and obtained analytically apparently correspond to "laminate" structures. These will be examined in great detail in Chapter 9 in connection with the rank-one convexity condition for the vector case. These structures are a special case of gradient parametrized measures that admit a nice analytical treatment. They can be characterized by means of Jensen's inequality for rank-one convex functions.

3. The two-well problem

An interesting way of clarifying precise some of the ideas described in the previous sections and of demonstrating how complicated some issues can be concerning the full description of microstructures is to study the two-well problem where we assume that the zero set of the energy density φ, $\{\varphi = 0\}$, is the union of two disjoint wells. We restrict attention to dimension 2 so that $\mathbf{M}$ denotes throughout this section the set of 2×2 matrices.

The problem we want to address is the search for minimizers of the variational problem

$$I(u) = \int_\Omega \varphi(\nabla u(x))\, dx, \quad u = u_0 \text{ on } \partial\Omega,$$

where u_0 is some prescribed Lipschitz deformation with finite energy $I(u_0) < \infty$. Our main assumption here is that the free energy density φ is nonnegative, $\varphi(Y) = +\infty$ if $\det Y \leq 0$ and

$$K = \{\varphi = 0\} = SO(2)F_1 \cup SO(2)F_2, \quad \det F_i > 0.$$

We have already noted the fact that under these assumptions φ cannot be quasiconvex and we cannot expect classical minimizers. For this reason we concentrate on the generalized equivalent variational principle

$$\tilde{I}(\nu) = \int_\Omega \int_{\mathbf{M}} \varphi(A)\, d\nu_x(A)\, dx$$

where $\nu = \{\nu_x\}_{x\in\Omega}$ must be a gradient parametrized measure (a microstructure) with compact support (we restrict attention to $p = \infty$ in the context of Chapters 4 and 8) and $u - u_0 \in W_0^{1,\infty}(\Omega)$ where

$$\nabla u(x) = \int_{\mathbf{M}} A\, d\nu_x(A).$$

We are interested in finding *stress-free microstructures*: $\tilde{I}(\nu) = 0$. Obviously this is equivalent to actually having $\operatorname{supp}(\nu_x) \subset K$ and this condition in turn imposes restrictions on the set K itself, and on the possible boundary values u_0 that may support nontrivial, stress-free microstructures. We take the term microstructure here as equivalent to gradient parametrized measure.

We would like to draw some conclusions regarding three issues raised in the above paragraph:

1. Conditions on the two wells to ensure the existence of nontrivial, stress-free microstructures.
2. Affine boundary values that may support such microstructures.
3. Examples of stress-free microstructures.

Underlying our analysis is the need to understand the convex hull of the set $SO(2)$. Indeed, it consists of all matrices P of the form

$$P = \begin{pmatrix} \alpha & \beta \\ -\beta & \alpha \end{pmatrix}, \qquad \det P = \alpha^2 + \beta^2 \leq 1, \tag{5–5}$$

which is elementary to verify. Furthermore, if μ is a probability measure on $SO(2)$,

$$P = \int_{SO(2)} Q\, d\mu(Q)$$

is in the convex hull of $SO(2)$. If $\det P = 1$ then $P \in SO(2)$ and $\mu = \delta_P$.

The main tool in deriving necessary conditions in this context is the minor relation

$$\det\left(\int_K A\, d\nu(A)\right) = \int_K \det A\, d\nu(A)$$

which should be valid whenever ν is a gradient parametrized measure. This equality is also true for $\psi(A) = \det(A - F)$ for a fixed matrix F because ψ is also a weak continuous function. This fact can also be proved by using the formula that follows which will play a role in some proofs. It is only valid for 2×2 matrices

$$\det(A - B) = \det A - (\operatorname{adj} A)^T \cdot B + \det B, \tag{5–6}$$

where $A \operatorname{adj} A = \det A\, \mathbf{1}$. Notice that $(\operatorname{adj} A)^T \cdot B$ is a linear function on the entries of A.

Before going any further, we would like to consider if it is possible to have nontrivial gradient parametrized measures supported in a single well $SO(2)F$. This is impossible due to the weak continuity of the determinant: if ν is a gradient parametrized measure supported in $SO(2)F$,

$$\det\left(\int_{SO(2)F} A\, d\nu(A)\right) = \int_{SO(2)F} \det A\, d\nu(A).$$

The right-hand side is $\det F$ and the left-hand side can be written $\det(PF)$ where $P \in \operatorname{co}(SO(2))$. Hence $\det P = 1$ and by the observation above this implies that ν has to be trivial, a delta measure.

A second easy preliminary step is to consider gradient parametrized measures supported in just two matrices, F_1 and F_2. In this case any probability measure can be written

$$\nu = (1 - \lambda)\delta_{F_1} + \lambda\delta_{F_2}, \quad \lambda \in [0, 1].$$

Again, by the weak continuity of the determinant,

$$\det\left((1 - \lambda)F_1 + \lambda F_2\right) = (1 - \lambda)\det(F_1) + \lambda \det(F_2).$$

The left-hand side can be decomposed as

$$(1 - \lambda)\det(F_1) + \lambda\det(F_2) - \lambda(1 - \lambda)\det(F_1 - F_2).$$

This formula is also valid only for 2×2 matrices. This condition clearly implies $\det(F_1 - F_2) = 0$ and thus $F_1 - F_2$ must be a rank-one matrix. Otherwise ν must be a Dirac mass.

We now get fully into the two-well problem, and treat in succession the three issues mentioned above.

1. Our main result concerning restrictions on the set of two wells is the following. We say that the wells $SO(2)F_1$ and $SO(2)F_2$ are incompatible if $F_1 - QF_2$ is never a rank-one matrix for all rotations Q.

Theorem 5.1 *Let ν be a homogeneous gradient parametrized measure with*

$$\operatorname{supp}\nu \subset SO(2)F_1 \cup SO(2)F_2, \qquad \det F_i > 0, \quad i = 1, 2.$$

If the wells $SO(2)F_1$ and $SO(2)F_2$ are not compatible, $\nu = \delta_{QH_i}$ is a Dirac mass.

A crucial technical fact in the proof is the next lemma.

Lemma 5.2 *Let A be a matrix such that $\det(A - Q) > 0$ for all rotations $Q \in SO(2)$. Then $\det(A - P) > 0$ for every $P \in co(SO(2))$.*

Proof. Write

$$A = \begin{pmatrix} a_{11} & a_{12} \\ a_{21} & a_{22} \end{pmatrix}, \quad Q = \begin{pmatrix} \alpha & \beta \\ -\beta & \alpha \end{pmatrix}, \ \alpha^2 + \beta^2 = 1.$$

After some algebra

$$\det(A - Q) = \left(\alpha - \frac{a_{11} + a_{22}}{2}\right)^2 + \left(\beta - \frac{a_{21} - a_{12}}{2}\right)^2 - \left(\frac{a_{11} - a_{22}}{2}\right)^2 - \left(\frac{a_{21} + a_{12}}{2}\right)^2.$$

If $\det(A - Q) > 0$ for all (α, β) in the unit circle, this means that the unit circle does not meet the circle centered at

$$\left(\frac{a_{11} + a_{22}}{2}, \frac{a_{21} - a_{12}}{2}\right)$$

with radius

$$\sqrt{\left(\frac{a_{11} - a_{22}}{2}\right)^2 + \left(\frac{a_{21} + a_{12}}{2}\right)^2}.$$

By continuity, this last circle does not meet the solid unit circle either. This is the conclusion of the lemma. ■

Proof of Theorem 5.1. Set

$$\nu = (1 - \lambda)\nu^1 + \lambda\nu^2, \quad \operatorname{supp}(\nu^i) \subset SO(2)F_i, \quad i = 1, 2,$$

$$P_i = \left(\int_{SO(2)F_i} A \, d\nu^i(A)\right) F_i^{-1} \in \operatorname{co}(SO(2)), \quad i = 1, 2,$$

$$F = \int_K A \, d\nu(A) = (1 - \lambda)P_1F_1 + \lambda P_2F_2.$$

Consider the weak continuous function

$$\psi(A) = \det(A - P_2F_2).$$

On the one hand, by direct substitution

$$\psi(F) = (1-\lambda)^2 \det(P_1F_1 - P_2F_2). \tag{5–7}$$

Due to the weak continuity, and by (5–6)

$$\begin{aligned}\psi(F) =&(1-\lambda)\int_{SO(2)F_1} \det(A - F_2P_2)\,d\nu^1(A)\\ &+\lambda\int_{SO(2)F_2} \det(A - F_2P_2)\,d\nu^2(A)\\ =&(1-\lambda)\left(\det(F_1) + \det(P_2F_2) - (\operatorname{adj}(P_1F_1))^T\cdot(P_2F_2)\right)\\ &+\lambda\left(\det(F_2) + \det(P_2F_2) - (\operatorname{adj}(P_2F_2))^T\cdot(P_2F_2)\right)\\ =&(1-\lambda)\left(\det(F_1) - \det(P_1F_1) + \det(P_1F_1 - P_2F_2)\right)\\ &+\lambda\left(\det(F_2) - \det(P_2F_2)\right).\end{aligned}$$

Therefore we obtain the equality

$$\begin{aligned}(1-\lambda)^2\det(P_1F_1 - P_2F_2) =&(1-\lambda)(1-\det(P_1))\det(F_1)\\ &+\lambda(1-\det(P_2))\det(F_2)\\ &+(1-\lambda)\det(P_1F_1 - P_2F_2),\end{aligned}$$

or

$$\begin{aligned}(1-\lambda)(1-\det(P_1))\det(F_1) + \lambda(1-\det(P_2))\det(F_2)&\\ +\lambda(1-\lambda)\det(P_1F_1 - P_2F_2) = 0.&\end{aligned} \tag{5–8}$$

Assume that $\lambda \in (0,1)$ and the wells are incompatible, so that $\det(RF_1 - QF_2) > 0$ for all rotations Q and R. Multiplying by F_2^{-1} to the right and letting $A = RF_1F_2^{-1}$ we have $\det(A - Q) > 0$ for all rotations Q. By Lemma 5.2, $\det(A-P) > 0$ for all P in the convex hull. This is equivalent to $\det(RF_1 - PF_2) > 0$ for all such P. In particular $\det(RF_1 - P_2F_2) > 0$ for all rotations R. Therefore

$$\int_{SO(2)F_1} \det(A - P_2F_2)\,d\nu^1(A) > 0. \tag{5–9}$$

By the formula used above

$$\begin{aligned}\psi(F) =&(1-\lambda)\int_{SO(2)F_1} \det(A - F_2P_2)\,d\nu^1(A)\\ &+\lambda\int_{SO(2)F_2} \det(A - F_2P_2)\,d\nu^2(A).\end{aligned}$$

The first term of the right-hand side is positive by (5–9) and the second term, by the computations made earlier, is equal to

$$\lambda\,(1-\det(P_2))\det(F_2)$$

which is nonnegative (recall $1-\det(P_i)\geq 0$). Hence $\psi(F)>0$ and by (5–7), $\det(P_1F_1-P_2F_2)>0$. This is a clear contradiction of (5–8) because the sum of three nonnegative terms vanishes only if each one vanishes individually. The conclusion is that if the wells are incompatible, then either $\lambda=0$ or $\lambda=1$ and in this case the probability measure is trivial (case of one well). ∎

2. We would like to characterize the affine boundary conditions $u_0(x)=Fx$, $F\in\mathbf{M}$, that may support nontrivial, stress-free microstructures. We assume accordingly that the two wells are compatible. After an appropriate change of coordinates we can take

$$K=SO(2)F_0\cup SO(2)F_0^{-1},\quad F_0^{\pm 1}=\mathbf{1}\pm\delta e_1\otimes e_2,$$

where $\delta>0$ is a fixed parameter and e_i is the canonical basis for $\mathbf{R}^2$. If ν is a homogeneous gradient parametrized measure, we write

$$\nu=(1-\lambda)\nu^1+\lambda\nu^2,$$

and hence

$$F=\int_K A\,d\nu(A)=(1-\lambda)P_1F_0+\lambda P_2F_0^{-1},$$

where $P_i\in\operatorname{co}(SO(2))$ and

$$P_i=\int_{SO(2)} A\,d\nu^i(A)F_i^{-1}=\begin{pmatrix}\alpha_i & \beta_i\\ -\beta_i & \alpha_i\end{pmatrix},\quad \alpha_i^2+\beta_i^2\leq 1.$$

We have kept the notation $F_1=F_0$, $F_2=F_0^{-1}$ for convenience. Placing these expressions into F,

$$F=(1-\lambda)\begin{pmatrix}\alpha_1 & \alpha_1\delta+\beta_1\\ -\beta_1 & -\beta_1\delta+\alpha_1\end{pmatrix}+\lambda\begin{pmatrix}\alpha_2 & -\alpha_2\delta+\beta_2\\ -\beta_2 & \beta_2\delta+\alpha_2\end{pmatrix},$$

and for $C=F^TF$, the Cauchy-Green tensor, write

$$C=F^TF=\begin{pmatrix}c_{11} & c_{12}\\ c_{21} & c_{22}\end{pmatrix},\quad F=\begin{pmatrix}F^{(1)} & F^{(2)}\end{pmatrix}.$$

Then we have the inequalities

$$\begin{aligned}
c_{11}=\left|F^{(1)}\right|^2 &\leq(1-\lambda)\left(\alpha_1^2+\beta_1^2\right)+\lambda\left(\alpha_2^2+\beta_2^2\right)\leq 1,\\
c_{22}=\left|F^{(2)}\right|^2 &\leq(1-\lambda)\left|(\beta_1,\alpha_1)+\delta(\alpha_1,-\beta_1)\right|^2\\
&\quad+\lambda\left|(\beta_2,\alpha_2)+\delta(-\alpha_2,\beta_2)\right|^2\\
&=(1-\lambda)(1+\delta^2)+\lambda(1+\delta^2)\\
&=(1+\delta^2).
\end{aligned}$$

On the other hand by the weak continuity of det,

$$\det F = \int_K \det A \, d\nu(A) = 1$$

so that

$$1 = \det C = c_{11}c_{22} - c_{12}^2,$$

and consequently

$$c_{11}c_{22} \geq 1.$$

In the c_{11}-c_{22} plane we have found the constraints

$$c_{11} \leq 1, \quad c_{22} \leq (1+\delta^2), \quad c_{11}c_{22} \geq 1.$$

These determine a region D easy to draw. The question is: does every point in D come from the Cauchy- Green tensor corresponding to a gradient parametrized measure ν supported in K? The answer is yes. To understand this we need to review briefly how laminates supported in four matrices can be easily constructed. For a complete discussion on laminates and gradient parametrized measures, refer to Chapter 9.

With four matrices, A, B, C, D, the compatibility conditions we need in order to have a laminate are

$$\begin{gathered}\operatorname{rank}(A-B) = 1, \quad \operatorname{rank}(C-D) = 1, \\ \operatorname{rank}\ ((\lambda A + (1-\lambda)B) - (\sigma C + (1-\sigma)D)) = 1,\end{gathered}$$

for some $\lambda, \sigma \in (0,1)$. In this case, any convex combination of $\lambda\delta_A + (1-\lambda)\delta_B$ and $\sigma\delta_C + (1-\sigma)\delta_D$ will be a gradient parametrized measure (a laminate), using again the idea of layers within layers to find the corresponding sequence of gradients (Chapter 9).

Let ν be the laminate supported in K

$$\nu = (1-\lambda)\delta_{F_0} + \lambda\delta_{F_0^{-1}},$$

(F_0 and F_0^{-1} are rank-one related). For this ν,

$$F = \begin{pmatrix} 1 & \delta - 2\lambda\delta \\ 0 & 1 \end{pmatrix},$$

and the corresponding Cauchy-Green tensor

$$C = F^T F = \begin{pmatrix} 1 & \delta - 2\lambda\delta \\ \delta - 2\lambda\delta & 1 + \delta^2(1-2\lambda)^2 \end{pmatrix}.$$

As λ moves from 0 to 1, $c_{22} = 1 + \delta^2(1-2\lambda)^2$ goes down from $1+\delta^2$ to 1 and then back to $1+\delta^2$, while c_{11} stays constant at 1.

There is another matrix $Q_\delta \in SO(2)$ with the property that $Q_\delta F_0$ is rank-one related to F_0^{-1}. Namely, after some computations,

$$Q_\delta = \frac{1}{1+\delta^2}\begin{pmatrix} 1-\delta^2 & -2\delta \\ 2\delta & 1-\delta^2 \end{pmatrix}. \tag{5–10}$$

The matrix $Q_\delta F_0$ is called the reciprocal twin of F_0^{-1}. Thus we may consider the laminate

$$\tilde{\nu} = (1-\lambda)\delta_{Q_\delta F_0} + \lambda \delta_{F_0^{-1}}$$

and find

$$\tilde{F} = (1-\lambda)Q_\delta F_0 + \lambda F_0^{-1}.$$

In this case one obtains

$$c_{11} = \frac{1}{1+\delta^2}\left(1+\delta^2(2\lambda-1)^2\right), \quad c_{22} = 1+\delta^2,$$

so that as λ runs through $[0,1]$, c_{22} is fixed at $1+\delta^2$ but c_{11} goes from 1 to $1/(1+\delta^2)$ and back to 1.

These very same computations show that for a given $\lambda \in [0,1]$ and $Q_{(\lambda)} = Q_{\delta(1-2\lambda)}$, given by (5–10) with $\delta(1-2\lambda)$ replacing δ, the matrix

$$Q_{(\lambda)}\left((1-\lambda)F_0 + \lambda F_0^{-1}\right)$$

is the reciprocal twin of

$$\left((1-\lambda)F_0 + \lambda F_0^{-1}\right)^{-1} = (1-\lambda)F_0^{-1} + \lambda F_0,$$

because $(1-\lambda)F_0 + \lambda F_0^{-1}$ is a matrix of the same type as F_0. For

$$F = (1-\sigma)Q_{(\lambda)}\left((1-\lambda)F_0 + \lambda F_0^{-1}\right) + \sigma\left((1-\lambda)F_0^{-1} + \lambda F_0\right),$$

we reach eventually every point in D as $(\sigma,\lambda) \in [0,1]\times[0,1]$, for λ lets us move up and down and σ from left to right. This F corresponds to the measure

$$\begin{aligned}\nu = (1-\sigma)(1-\lambda)\delta_{Q_{(\lambda)}F_0} + (1-\sigma)\lambda\delta_{Q_{(\lambda)}F_0^{-1}} \\ +\sigma(1-\lambda)\delta_{F_0^{-1}} + \sigma\lambda\delta_{F_0},\end{aligned}$$

where

$$Q_{(\lambda)} = \frac{1}{1+\delta^2(1-2\lambda)^2}\begin{pmatrix} 1-\delta^2(1-2\lambda)^2 & -2\delta(1-2\lambda) \\ 2\delta(1-2\lambda) & 1-\delta^2(1-2\lambda)^2 \end{pmatrix}. \tag{5–11}$$

This probability measure is a laminate because the rotation $Q_{(\lambda)}$ was so determined.

3. The next step is to study, for each possible F whose Cauchy-Green tensor lies in D, the set of gradient parametrized measures supported in the two wells with such an underlying deformation, or at least to say something about the structure or the complexity of that set. As we will shortly see, this is a much harder problem that cannot be solved completely except for some special matrices.

Suppose that $\nu = \{\nu_x\}_{x\in\Omega}$ is a nonhomogeneous, gradient parametrized measure supported in K where we take again $F_1 = F_0$, $F_2 = F_0^{-1}$: $\nu_x = (1-\lambda(x))\,\nu_x^1 + \lambda(x)\nu_x^2$. Denote by $y(x)$ the deformation underlying ν, that is,

$$\begin{aligned}\nabla y(x) &= \int_{\mathbf{M}} A\, d\nu_x \\ &= (1-\lambda(x))\int_{SO(2)F_1} Q\, d\nu_x^1(Q) + \lambda(x)\int_{SO(2)F_2} Q\, d\nu_x^2(Q) \\ &= (1-\lambda(x))\, P_1(x)\, F_1 + \lambda(x) P_2(x)\, F_2\end{aligned} \tag{5–12}$$

where

$$P_i(x) = \int_{SO(2)F_i} Q\, d\nu_x^i(Q)\, F_i^{-1}, \quad i = 1, 2,$$

belong to the convex hull of $SO(2)$. We have the following uniqueness result.

Theorem 5.3 *Suppose that $y(x)$ satisfies*

$$y(x) = Fx = (1-\theta)F_1 x + \theta F_2 x, \quad x \in \partial\Omega,$$

for some θ, $0 < \theta < 1$. Then

$$\nu_x = (1-\theta)\delta_{F_1} + \theta\delta_{F_2}, \quad \textit{for } x \in \Omega.$$

Proof. Assuming that $|\Omega| = 1$, by the divergence theorem,

$$\begin{aligned}F &= \int_\Omega \nabla y(x)\, dx \\ &= \int_\Omega (1-\lambda(x)) \int_{SO(2)F_1} Q\, d\nu_x^1\, dx + \int_\Omega \lambda(x) \int_{SO(2)F_2} Q\, d\nu_x^2\, dx \\ &= (1-\lambda^*) \int_{\Omega\times SO(2)F_1} Q \frac{(1-\lambda(x))}{1-\lambda^*}\, d\nu_x^1\, dx + \lambda^* \int_{\Omega\times SO(2)F_2} Q \frac{\lambda(x)}{\lambda^*}\, d\nu_x^2\, dx \\ &= (1-\lambda^*)\, M_1 F_1 + \lambda^* M_2 F_2,\end{aligned}$$

where λ^* is the average of λ over Ω. Now

$$\frac{(1-\lambda(x))}{1-\lambda^*}\, d\nu_x^1\, dx \quad \text{and} \quad \frac{\lambda(x)}{\lambda^*}\, d\nu_x^2\, dx$$

are probability measures, and hence reduce to Dirac masses if the M_i are rotations. Furthermore, the M_i are averages of rotations, and hence lie in the convex hull of $SO(2)$.

We now have the equation

$$(1-\theta)F_1 + \theta F_2 = (1-\lambda^*)\,M_1F_1 + \lambda^* M_2F_2.$$

Multiplying to the right by $F_1^{-1} = F_0^{-1} = F_2$,

$$(1-\theta)\mathbf{1} + \theta H = (1-\lambda^*)\,M_1 + \lambda^* M_2 H, \tag{5–13}$$

where $H = \left(F_0^{-1}\right)^2 = \mathbf{1} + te_1 \otimes e_2$, $t = -2\delta$. Say that

$$M_i = \begin{pmatrix} \alpha_i & \beta_i \\ -\beta_i & \alpha_i \end{pmatrix}, \quad \alpha_i^2 + \beta_i^2 \leq 1, \ i = 1, 2.$$

Then

$$\begin{pmatrix} 1 & \theta t \\ 0 & 1 \end{pmatrix} = (1-\lambda^*)\begin{pmatrix} \alpha_1 & \beta_1 \\ -\beta_1 & \alpha_1 \end{pmatrix} + \lambda^* \begin{pmatrix} \alpha_2 & \alpha_2 t + \beta_2 \\ -\beta_2 & -\beta_2 t + \alpha_2 \end{pmatrix}.$$

Now $|\alpha_i| \leq 1$, and

$$1 = (1-\lambda^*)\,\alpha_1 + \lambda^* \alpha_2$$

implies $\alpha_i = 1$. Next

$$1 = (1-\lambda^*)\,\alpha_1 + \lambda^*(\alpha_2 - \beta_2 t) = 1 - \lambda^* \beta_2 t$$

implies $\beta_2 = 0$. Finally,

$$0 = (1-\lambda^*)\,\beta_1 + \lambda^* \beta_2 = (1-\lambda^*)\,\beta_1$$

can only happen if $\beta_1 = 0$, and likewise

$$\theta t = (1-\lambda^*)\,\beta_1 + \lambda^*\left(\beta_2 + \alpha_2 t\right) = \lambda^* t$$

implies $\theta = \lambda^*$. Consequently the matrices $M_i = \mathbf{1}$, $\nu_x^i = \delta_1$, $i = 1, 2$, and

$$\nu_x = (1 - \lambda(x))\,\delta_{F_1} + \lambda(x)\delta_{F_2}.$$

We need now to show that $\lambda(x)$ is actually a constant function. First, using the mixed second partial derivatives in (5–12) with $P_i = \mathbf{1}$, we conclude that $\lambda(x)$ is a function of x_2 alone. Then

$$y(x) = x + f(x_2)e_1.$$

Applying the boundary condition, we see that $f(x_2) = \theta t x_2$, and going back to (5–12), we obtain $\lambda(x) \equiv \theta$. ■

This uniqueness result is very special. Indeed for most of the matrices that may support nontrivial microstructures such uniqueness fails drastically: there even exist continuously distributed gradient parametrized measures supported in the two wells. The construction that follows is based on two of the main facts shown in Chapter 8 and stated in Chapter 1:

i) the process of averaging when we have affine boundary values, Theorem 8.1; and
ii) the decoupling in rank-one compatibility and oscillatory properties of non-homogeneous gradient parametrized measures, the characterization theorem, Theorem 8.16.

Let us take $\Omega = [0,1] \times [0,1]$, and let $y : \Omega \to \mathbf{R}^2$ be a deformation with some affine boundary condition. Assume that we can actually find y with the property that $F(x) = \nabla y(x)$ admits the decomposition

$$\begin{aligned} F(x) =& P(x)\left[\mu(x)\lambda(x)Q(\lambda(x)) + (1-\mu(x))\,(1-\lambda(x))\,\mathbf{1}\right] F_0 \\ &+ P(x)\left[\mu(x)\,(1-\lambda(x))\,Q(\lambda(x)) + (1-\mu(x))\,\lambda(x)\mathbf{1}\right] F_0^{-1}, \end{aligned} \tag{5–14}$$

where we are using the same notation as in the previous section, $F_0^{\pm 1} = \mathbf{1} \pm \delta e_1 \otimes e_2$, $\delta > 0$, λ and μ are nonconstant, continuous functions with values in $[0,1]$, $P : \Omega \to SO(2)$ and $Q(\lambda(x)) \in SO(2)$ given by (5–11) is such that

$$\det\left\{Q(\lambda(x))\left[\lambda(x)F_0 + (1-\lambda(x))\,F_0^{-1}\right] - \left[\lambda(x)F_0^{-1} + (1-\lambda(x))\,F_0\right]\right\} = 0.$$

We claim that the family of probability measures $\nu = \{\nu_x\}_{x\in\Omega}$ given by

$$\begin{aligned} \nu_x =& \left[\mu(x)\lambda(x)\delta_{P(x)Q(\lambda(x))F_0} + (1-\mu(x))\,(1-\lambda(x))\,\delta_{P(x)F_0}\right] \\ &+ \left[\mu(x)\,(1-\lambda(x))\,\delta_{P(x)Q(\lambda(x))F_0^{-1}} + (1-\mu(x))\,\lambda(x)\delta_{P(x)F_0^{-1}}\right], \end{aligned}$$

is a gradient parametrized measure. This is a direct consequence of Theorem 8.16 above since by the preceding discussion, each ν_x is a laminate supported in K. Therefore under the assumption (5–14) our claim is true.

Let us look at the average of such ν, $\overline{\nu}$. According to the average formula, for a continuous function ψ,

$$\begin{aligned} \int_K \psi(A)\,d\overline{\nu} =& \int_\Omega \int_K \psi(A)\,d\nu_x(A)\,dx \\ =& \int_\Omega \left[\mu(x)\lambda(x)\psi\left(P(x)Q(\lambda(x))F_0\right)\right. \\ &+ (1-\mu(x))\,(1-\lambda(x))\,\psi\left(P(x)F_0\right) \\ &+ \mu(x)\,(1-\lambda(x))\,\psi\left(P(x)Q(\lambda(x))F_0^{-1}\right) \\ &+ \left.(1-\mu(x))\,\lambda(x)\psi\left(P(x)F_0^{-1}\right)\right]\,dx. \end{aligned}$$

If $\lambda(x)$ is a continuous, nonconstant function, either $\{P(x)Q(\lambda(x))F_0\}_{x\in\Omega}$ or $\{P(x)F_0\}_{x\in\Omega}$ is a continuous distribution on the well corresponding to F_0. Since the density functions $\mu(x)\lambda(x)$ and $(1-\mu(x))(1-\lambda(x))$ are both nonnegative and nonconstant, the equality

$$\int_K \psi(A)\, d\bar{\nu}_1(A) = \int_\Omega [\mu(x)\lambda(x)\psi(P(x)Q(\lambda(x))F_0) \\ + (1-\mu(x))(1-\lambda(x))\psi(P(x)F_0)]\, dx$$

asserts that $\overline{\nu}_1$ is a continuous distribution on $SO(2)F_0$. The same argument is valid for the well $SO(2)F_0^{-1}$.

Let us find a function $y : \Omega = [0,1]^2 \to \mathbf{R}^2$ for which the decomposition (5–14) can be achieved. We know that this decomposition is possible if for $C = F^T F = \nabla y^T \nabla y$, we have the constraints

$$c_{11}c_{22} - c_{12}^2 = 1, \\ 1 \le c_{22} \le 1 + \delta^2, \\ \frac{1}{1+\delta^2} \le c_{11} \le 1,$$

where as before

$$C = \begin{pmatrix} c_{11} & c_{12} \\ c_{12} & c_{22} \end{pmatrix}.$$

First of all, a map φ of type

$$(r,\theta) \to (r, \theta + \epsilon\, \xi(r)), \\ \xi \in \mathcal{C}^1([0,1]), \quad \xi(0) = \xi(1) = 0, \quad 0 \le r \le 1, \quad \epsilon > 0,$$

in polar coordinates, has the properties:

i) $\det \nabla\varphi = 1$,

ii) $\varphi|_{r=1} = \mathrm{id}$.

In rectangular coordinates $\varphi = (\varphi_1(x_1,x_2), \varphi_2(x_1,x_2))$, and it is elementary to find

$$\varphi_1 = x_1 \cos(\epsilon\, \xi(r)) - x_2 \sin(\epsilon\, \xi(r)), \\ \varphi_2 = x_2 \cos(\epsilon\, \xi(r)) + x_1 \sin(\epsilon\, \xi(r)).$$

Direct computation yields

$$\left(\frac{\partial \varphi_1}{\partial x_1}\right)^2 + \left(\frac{\partial \varphi_2}{\partial x_1}\right)^2 = 1 + x_1^2 (\epsilon \xi'(r))^2 - 2\epsilon\xi'(r)\frac{x_1 x_2}{r}, \\ r = \sqrt{x_1^2 + x_2^2},$$

and something similar for

$$\left(\frac{\partial \varphi_1}{\partial x_2}\right)^2 + \left(\frac{\partial \varphi_2}{\partial x_2}\right)^2.$$

The point is that these two expressions, that represent the diagonal of the Cauchy-Green tensor of the deformation φ, are nonconstant in any subdomain for almost any choice of ξ (take for instance $\xi(r) = r(1-r)$). According to our discussion, this in turn ensures that λ and σ are nonconstant functions.

Given $\alpha \in (0,1)$, consider now φ_α, a variant of φ itself,

$$(r,\theta) \to \left(r, \theta + \alpha\,\epsilon\,\xi\left(\frac{r}{\alpha}\right)\right), \quad 0 \le r \le \alpha,$$

and extend it by the identity to the box

$$\left[\frac{-\alpha}{2}, \frac{-\alpha}{2}\right] \times \left[\frac{-1}{2\alpha}, \frac{-1}{2\alpha}\right].$$

After a translation, let u_α denote the corresponding map

$$u_\alpha : \Omega_\alpha = [0,\alpha] \times \left[0, \frac{1}{\alpha}\right] \to \Omega_\alpha,$$
$$\det \nabla u_\alpha = 1,$$
$$u_\alpha|_{\partial\Omega_\alpha} = \mathrm{id}.$$

When $\epsilon = 0$,

$$\left(\frac{\partial u_\alpha^1}{\partial x_i}\right)^2 + \left(\frac{\partial u_\alpha^2}{\partial x_i}\right)^2 = 1, \quad i = 1,2.$$

Therefore we can fix $\epsilon(\alpha) > 0$, sufficiently small so that

$$\frac{1}{\sqrt{1+\delta^2}} \le \left(\frac{\partial u_\alpha^1}{\partial x_i}\right)^2 + \left(\frac{\partial u_\alpha^2}{\partial x_i}\right)^2 \le \sqrt{1+\delta^2}, \quad i = 1,2.$$

(Recall that $F_0^{\pm} = \mathbf{1} \pm \delta e_1 \otimes e_2$, $\delta > 0$.) Finally, let y be

$$y = u_\alpha \cdot H_\alpha : \Omega = [0,1]^2 \to \Omega_\alpha,$$

for $H_\alpha = \alpha e_1 \otimes e_1 + \frac{1}{\alpha} e_2 \otimes e_2$, and $\alpha = \left(1+\delta^2\right)^{-1/4} < 1$. Clearly $\det \nabla y = 1$ and

$$c_{11} = \left(\frac{\partial y^1}{\partial x_1}\right)^2 + \left(\frac{\partial y^2}{\partial x_1}\right)^2 = \alpha^2 \left[\left(\frac{\partial u_\alpha^1}{\partial x_1}\right)^2 + \left(\frac{\partial u_\alpha^2}{\partial x_1}\right)^2\right],$$

$$c_{22} = \left(\frac{\partial y^1}{\partial x_2}\right)^2 + \left(\frac{\partial y^2}{\partial x_2}\right)^2 = \frac{1}{\alpha^2} \left[\left(\frac{\partial u_\alpha^1}{\partial x_2}\right)^2 + \left(\frac{\partial u_\alpha^2}{\partial x_2}\right)^2\right],$$

so that the following inequalities are valid

$$\begin{aligned}
c_{11} &\le (1+\delta^2)^{-1/2}(1+\delta^2)^{1/2} = 1,\\
c_{11} &\ge \alpha^2 \frac{1}{\sqrt{1+\delta^2}} = \frac{1}{1+\delta^2},\\
c_{22} &\le (1+\delta^2)^{1/2}(1+\delta^2)^{1/2} = 1+\delta^2,\\
c_{22} &\ge \frac{1}{\alpha^2}\frac{1}{\sqrt{1+\delta^2}} = 1.
\end{aligned}$$

Therefore ∇y admits the claimed decomposition. Moreover, in the subdomain $H_\alpha^{-1}\left(B_{\alpha/2}\left(\frac{\alpha}{2},\frac{1}{2\alpha}\right)\right) \subset \Omega$, c_{11} and c_{22} are nonconstant by the computations made earlier.

We have obtained a homogeneous, continuously distributed gradient parametrized measure supported on the set of the two wells.

4. An example in micromagnetics

Once we know that the variational principle of micromagnetics as explained in the introduction does not lend itself directly to study by the direct method and minimizing sequences may develop oscillations, we introduce the notion of measure-valued magnetization. We are willing to accept a measure-valued solution in the sense that the oscillations described by minimizing sequences take place in so fine a scale that we only care about the states that participate in the oscillations and the relative volume fractions of the regions in which such states occur. These two pieces of information are contained in the parametrized measure through the support and the weight for each state in the support, respectively. Therefore we would like to let parametrized measures $\nu = \{\nu_x\}_{x\in\Omega}$ compete in the variational principle (5–4). Let us consider a sequence of magnetizations, $\{m^k\}$, and let $\nu = \{\nu_x\}_{x\in\Omega}$ be its associated parametrized measure. Because m^k takes values on the unit sphere $\mathcal{S} = \{y \in \mathbf{R}^N : |y| = 1\}$, it is clear that the support of ν_x is contained in $\mathcal{S}$ for a.e. $x \in \Omega$. Moreover,

$$\lim_{k\to\infty}\int_\Omega \varphi(m^k)\,dx = \int_\Omega\int_{\mathbf{R}^N} \varphi(\lambda)\,d\nu_x(\lambda)\,dx.$$

On the other hand, if

$$m(x) = \int_{\mathbf{R}^N} \lambda\,d\nu_x(\lambda), \quad x \in \Omega,$$

then,

$$m^k \overset{*}{\rightharpoonup} m \quad \text{in } L^\infty(\Omega),$$

and the limit of the interaction energy is

$$-\int_{\Omega} H \cdot m\, dx.$$

The magnetostatic energy, however, presents a problem when trying to identify the limit energy in terms of the parametrized measure, because the relationship between the potential u and the magnetization m is given through the differential constraint $\operatorname{div}(-\nabla u + m\chi_\Omega) = 0$. The clue to understanding this passage to the limit for the magnetostatic energy is the following fact.

Theorem 5.4 *For any sequence of magnetizations, $\{m^k\}$, such that $\{\operatorname{div} m^k\}$ is a compact set in $H^{-1}_{loc}(\mathbf{R}^N)$, we have*

$$u^k \to u \text{ (strongly) in } H^1(\mathbf{R}^N),$$

where

$$\begin{aligned} div(-\nabla u^k + m^k\chi_\Omega) = 0 \quad & in\ H^{-1}(\mathbf{R}^N), \\ m^k \overset{*}{\rightharpoonup} m \quad & in\ L^\infty(\Omega), \\ div(-\nabla u + m\chi_\Omega) = 0 \quad & in\ H^{-1}(\mathbf{R}^N). \end{aligned}$$

In particular, the limit magnetostatic energy is obtained through the weak limit m in the same way that it is obtained from a genuine magnetization, provided that $\{\operatorname{div} m^k\}$ is a compact set in $H^{-1}_{loc}(\mathbf{R}^N)$.

The proof of Theorem 5.4 is based on the Div-Curl lemma, a typical compensated compactness result.

Div-Curl Lemma 5.5 *Let Ω be a regular domain bounded or unbounded. Let $\{U_j\}$ converge weakly to U in $L^2(\Omega)$ and V_j to V in $L^\infty(\Omega)$. Suppose that $\{curl\, U_j\}$, $\{div\, V_j\}$ are compact in $H^{-1}_{loc}(\Omega)$. Then U_jV_j converges weaky in the sense of distributions to the product UV.*

For the proof of Theorem 5.4, apply the Div-Curl lemma to the sequences $\{\nabla u^k\}$ and $\{m^k\}$ for which the hypotheses of this lemma hold. Through a density argument we can obtain the convergence

$$\int_{\mathbf{R}^N} \nabla u^k\, m^k\, \chi_\Omega\, dx \to \int_{\mathbf{R}^N} \nabla u\, m\, \chi_\Omega\, dx.$$

Using the differential constraint

$$\int_{\mathbf{R}^N} \nabla u^k\, m^k\, \chi_\Omega\, dx = \int_{\mathbf{R}^N} \nabla u^k \nabla u^k\, dx$$
$$\int_{\mathbf{R}^N} \nabla u\, m\, \chi_\Omega\, dx = \int_{\mathbf{R}^N} \nabla u \nabla u\, dx.$$

This gives us the strong convergence of the gradients in $L^2(\Omega)$. Together with the weak convergence of the solution operator to the differential equation we get the desired strong convergence. ∎

The above considerations lead us to define a measure-valued magnetization as a family of probability measures $\nu = \{\nu_x\}_{x\in\Omega}$ whose support lies in the unit sphere $\mathcal{S}$ for a.e. $x \in \Omega$ and can be generated by a sequence of classical magnetizations, m^k, with $\{\operatorname{div} m^k\}$ a compact set in $H^{-1}_{loc}(\mathbf{R}^N)$. For such a generalized magnetization $\nu = \{\nu_x\}_{x\in\Omega}$ we define its total energy as

$$I(\nu) = \int_\Omega \int_{\mathbf{R}^N} \varphi(\lambda)\, d\nu_x(\lambda)\, dx + \frac{1}{2}\int_\Omega |\nabla u|^2\, dx - \int_\Omega H \cdot m\, dx,$$

where m is the first moment of ν and

$$\operatorname{div}(-\nabla u + m\chi_\Omega) = 0 \text{ in } H^{-1}(\mathbf{R}^N).$$

The importance of having $\{\operatorname{div} m^k\}$ be a compact set in $H^{-1}_{loc}(\mathbf{R}^N)$ consists in the fact

$$I(\nu) = \lim_{k\to\infty} I(m^k)$$

according to Theorem 5.4. If we drop the condition on the divergences, it is always true that if $\nu = \{\nu_x\}_{x\in\Omega}$ is the parametrized measure associated to a sequence of magnetizations $\{m^k\}$ then

$$I(\nu) \le \liminf_{k\to\infty} I(m^k), \tag{5–15}$$

using the weak continuity of the solution mapping for the differential equation and the convexity of the function $|\nabla u|^2$. The point is that the above inequality might be strict if we do not have some extra condition like the divergences being contained in a compact set in $H^{-1}_{loc}(\mathbf{R}^N)$. If this condition is not assumed the energy of the parametrized measure limit of $\{m^k\}$ might not be the limit of the energies of $\{m^k\}$ and the energy for ν would not have any physical relevance as indicated above. In this sense, we say that measure-valued magnetizations as defined are the ones that can be interpreted physically: they come from a sequence of classical magnetizations and their energy is precisely the limit of the energies of the magnetizations. If now $\overline{\mathcal{A}}$ stands for the set of all measure-valued magnetizations and $\mathcal{A}$, for the set of the classical ones, we have shown a relaxation result:

$$\inf_{\overline{\mathcal{A}}} I(\nu) = \inf_{\mathcal{A}} I(m).$$

What is remarkable is the fact that the additional constraint on the divergences does not restrict further the families of probability measures in $\overline{\mathcal{A}}$. This is a main consequence of our analysis of divergence-free parametrized measures in Chapter 10. Specifically Theorem 10.3 establishes that

$$\overline{\mathcal{A}} = \{\nu = \{\nu_x\}_{x\in\Omega} : \nu_x \text{ is a probability measure and } \operatorname{supp}\nu_x \subset \mathcal{S}, \text{ for a.e. } x \in \Omega\}.$$

We can reformulate the above conclusions in the context of Section 3, Chapter 1. Let $\mathcal{L}$ be

$$\mathcal{L} = \{\{(m_j, \nabla u_j)\} : |m_j| = 1, \text{ div } (-\nabla u_j + m_j \chi_\Omega) = 0\}.$$

We would like to characterize parametrized measures associated to sequences in $\mathcal{L}$. If we are willing to add the condition on the divergences of m_j to the definition of $\mathcal{L}$, the parametrized measures, $\mu = \{\mu_x\}_{x\in\Omega}$, associated to such sequences $\{(m_j, \nabla u_j)\}$ are

$$\mu_x = \nu_x \otimes \delta_{\nabla u(x)},$$

where $\nu = \{\nu_x\}_{x\in\Omega}$ is the parametrized measure corresponding to the magnetizations $\{m_j\}$ and

$$\text{div}\,(-\nabla u + m\chi_\Omega) = 0, \quad m(x) = \int_{\mathbf{R}^N} \lambda\, d\nu_x(\lambda).$$

Since the condition on the divergences does not restrict further ν, for our analysis we can stick to $\mathcal{L}$ incorporating this compactness condition on the divergences.

We would also like to understand relaxation in terms of the first moments of elements in $\overline{\mathcal{A}}$. Notice that these first moments are precisely the weak limits of sequences of magnetizations. Let

$$\mathcal{A}^1 = \left\{ m \in L^\infty(\Omega) : m(x) = \int_{\mathbf{R}^N} \lambda\, d\nu_x(\lambda), \nu = \{\nu_x\}_{x\in\Omega} \in \overline{\mathcal{A}} \right\}$$

and for $m \in \mathcal{A}^1$,

$$I^{**}(m) = \int_\Omega \varphi^{**}(m)\, dx + \frac{1}{2}\int_{\mathbf{R}^N} |\nabla u|^2\, dx - \int_\Omega H \cdot m\, dx, \tag{5–16}$$

where once again $\text{div}\,(-\nabla u + m\chi_\Omega) = 0$, and by φ^{**} we actually mean the convexification of

$$\hat{\varphi} = \begin{cases} \varphi, & \text{in } \mathcal{S}, \\ +\infty, & \text{in } \mathbf{R}^N - \mathcal{S}. \end{cases}$$

Theorem 5.6 *If φ is a continuous, nonnegative function defined on $\mathcal{S}$,*

$$\inf_{\mathcal{A}^1} I^{**}(m) = \inf_{\overline{\mathcal{A}}} I(\nu) = \inf_{\mathcal{A}} I(u).$$

Proof. Let d^1, $\overline{d}$ and d stand for those three infima, respectively. We have already proved that $\overline{d} = d$. We show here $d^1 = \overline{d}$. By convexity, it is clear that $d^1 \le \overline{d}$ since for a given $m \in \mathcal{A}^1$ corresponding to $\nu \in \overline{\mathcal{A}}$, the last two terms in the energies are the same, and we apply Jensen's inequality to the first term.

On the other hand, if $m \in \mathcal{A}^1$, by definition of φ^{**} we can find ν_x, a probability measure supported on $\mathcal{S}$ such that

$$\varphi^{**}(m(x)) = \int_{\mathcal{S}} \varphi(\lambda)\, d\nu_x(\lambda)$$

and

$$m(x) = \int_{\mathcal{S}} \lambda\, d\nu_x(\lambda),$$

for a.e. $x \in \Omega$. By Theorem 10.3, $\nu = \{\nu_x\}_{x\in\Omega}$ belongs to $\overline{\mathcal{A}}$ and comparing the energies of ν and m we conclude $\overline{d} \le d^1$. ■

Theorem 5.7 *For any φ as in the previous theorem, the infimum*

$$\inf_{\overline{\mathcal{A}}} I(\nu)$$

is attained.

Proof. Let $\{m^k\}$ be a minimizing sequence of classical magnetizations and $\nu = \{\nu_x\}_{x\in\Omega}$ its corresponding parametrized measure. Again by Theorem 10.3, ν is in $\overline{\mathcal{A}}$, and by the weak lower semicontinuity (5–15) and the previous result,

$$\overline{d} \le I(\nu) \le \liminf_{k\to\infty} I(m^k) = d^1 = \overline{d}.$$

Thus ν is a minimizer. ■

Corollary 5.8

i) If $\nu = \{\nu_x\}_{x\in\Omega}$ is a minimizer for I, its first moment m is a minimizer for I^{**} and

$$\varphi^{**}(m(x)) = \int_{\mathcal{S}} \varphi(\lambda)\, d\nu_x(\lambda). \tag{5–17}$$

ii) If m is a minimizer for I^{**}, and ν_x is such that (5–17) holds with

$$m(x) = \int_{\mathcal{S}} \lambda\, d\nu_x(\lambda),$$

for a.e. $x \in \Omega$, then $\nu = \{\nu_x\}_{x\in\Omega}$ is a minimizer for I.

As a final observation, in the absence of an external applied field ($H = 0$) and if φ is a continuous, nonnegative, even function with minimum value 0, we have

$$d^1 = \overline{d} = d = 0.$$

Notice that under these assumptions $m = 0$ is a minimizer for I^{**}, so that all the infima are 0. This is true in particular for two important situations: the uniaxial and the cubic cases when φ is assumed to attain the minimum value along one axis or three axes.

5. Bibliographical remarks

Nonconvex variational principles have been remarkably successful in the analysis of phase transitions and microstructure in crystalline materials. This energetic point of view was introduced in [34], after the pioneering work in [120]-[131]. Further analysis has followed and explored this perspective: [35], [42], [43], [44], [73], [77], [169], [170], [174], [181], [195], [196], [230]. We have already indicated some general books and other references in nonlinear elasticity. See also [139]. The computational issues raised by these models are quite appealing despite their great complexity. The number of works on this topic is increasing very rapidly. Most of them focus on multiple well problems in several situations under affine boundary conditions: [63], [75], [76], [81], [82]-[88], [162], [185], [202], [205], [206], [253], [263]. [207] is a comprehensive survey of all these developments.

The two-well problem has also been extensively studied. Most of the material in Section 3 is contained in [192]. [35], [44], [109], [230], [231], [244], [297], [319] are also relevant to this issue.

The model of micromagnetics presented here has been taken from [171], [172], [173]. References on the foundation of micromagnetics are [59], [60], [176]. See also [175]. Some other models that differ in several aspects are considered in [55], [275], [276]. The relaxation described here for the micromagnetics functional has been examined in [108] and [306] as well.

Simulation of magnetic behavior has been undertaken systematically in [182], [183], [184], [208], [209], [210].

Chapter 6
Parametrized Measures

1. Introduction

This chapter is devoted to general issues related to parametrized measures. For this reason it is of a technical nature. We start by establishing a rather general existence theorem that can be applied to most of the situations one encounters in practice. This existence theorem provides a representation of weak limits, when they exist, of any composition with the sequence under consideration in terms of the parametrized measure associated to such a sequence (or possibly to some subsequence). It is important to stress that this result does not guarantee in any way that the weak limit exists. This is something to be obtained independently. If we have weak limits then they can be represented by an appropriate integral against the parametrized measure. If the weak limit does not exist (because of concentration effects) then parametrized measures yield a different type of information that we have yet to fully understand.

Since this representation of weak limits in terms of the parametrized measure requires having first weak convergence in L^1 of the sequence involved, we dedicate some effort to understanding this weak convergence and how far it is from having uniform bounds in L^1. This issue leads us to consider biting convergence and to pay some attention to Chacon's biting lemma. It turns out that parametrized measures always provide biting limits. Whenever this biting convergence may be improved to weak convergence, parametrized measures will represent weak limits. We also give some sufficient conditions for this representation to be valid based on the De La Vallée-Poussin criterion of weak convergence and analyze the relationship between biting and weak convergence. We finally discuss how strong convergence becomes translated into the parametrized measure device.

This is a good place to point out that parametrized measures completely miss concentration effects. They have not been designed to account for this phenomenon. This fact will become clear as the reader moves through this chapter. The phenomenon is so dramatic that two different sequences, one exhibiting concentrations and the other one with the concentrations cut off, may share the same parametrized measure. For this reason, parametrized measures are especially well-suited to deal with bounded oscillations.

2. Existence theorem

It is important for us, in preparing for the existence theorem, to understand the weak convergence in $L^1(\Omega)$ of uniformly bounded sequences in the L^1-norm. The property of equiintegrability was recalled in Chapter 1. Remember that a sequence of L^1-functions, $\{f_j\}$, is said to be equiintegrable if for $\epsilon > 0$ given, one can find $\delta > 0$ (depending only on ϵ) such that

$$\int_E |f_j(x)|\, dx < \epsilon$$

for all j, if $|E| < \delta$. The following version of this property will prove to be useful.

Lemma 6.1 *Let $\{f_j\}$ be a bounded sequence in $L^1(\Omega)$,*

$$\|f_j\|_{L^1(\Omega)} \le C < \infty.$$

The sequence is weakly relatively compact in $L^1(\Omega)$ if and only if

$$\lim_{k\to\infty}\left(\sup_j \int_{\{|f_j|\ge k\}} |f_j|\, dx\right) = 0. \qquad (6\text{–}1)$$

Proof. Notice first that

$$\begin{aligned} |\{|f_j| \ge k\}| &\le \int_{\{|f_j|\ge k\}} 1\, dx \le \int_{\{|f_j|\ge k\}} \frac{|f_j|}{k}\, dx \\ &\le \int_\Omega \frac{|f_j|}{k}\, dx \le \frac{C}{k}, \end{aligned}$$

so that for k large, $|\{|f_j| \ge k\}|$ (a measurable set within bars stands for the Lebesgue measure of that set) is small uniformly in j, and by the equiintegrability property

$$\int_{\{|f_j|\ge k\}} |f_j|\, dx \le \epsilon,$$

for all j, and k large. Therefore the limit in (6–1) vanishes.

Conversely, if the above limit is zero and $\epsilon > 0$ is given, we can find k_0 such that

$$\int_{\{|f_j|\ge k\}} |f_j|\, dx \le \frac{\epsilon}{2},$$

for all j, and $k \geq k_0$. Set $\delta = \epsilon/(2k_0)$. If $E \subset \Omega$ and $|E| < \delta$,

$$\begin{aligned}\int_E |f_j|\,dx &= \int_{E\cap\{|f_j|\le k_0\}} |f_j|\,dx + \int_{E\cap\{|f_j|\ge k_0\}} |f_j|\,dx \\ &\le k_0\,|E| + \frac{\epsilon}{2} \\ &\le k_0\frac{\epsilon}{2k_0} + \frac{\epsilon}{2} \\ &= \epsilon,\end{aligned}$$

and the family of functions is equiintegrable. ■

In this section, we prove the basic existence theorem for parametrized measures. The hypotheses are pretty general so that this version can be applied to many different situations.

Theorem 6.2 *Let $\Omega \subset \mathbf{R}^N$ be a measurable set and let $z_j : \Omega \to \mathbf{R}^m$ be measurable functions such that*

$$\sup_j \int_\Omega g(|z_j|)\,dx < \infty, \tag{6–2}$$

where $g : [0,\infty) \to [0,\infty]$ is a continuous, nondecreasing function such that $\lim_{t\to\infty} g(t) = \infty$. There exists a subsequence, not relabeled, and a family of probability measures, $\nu = \{\nu_x\}_{x\in\Omega}$ (the associated parametrized measure) with the property that whenever the sequence $\{\psi(x, z_j(x))\}$ is weakly convergent in $L^1(\Omega)$ for any Carathéodory function $\psi(x,\lambda) : \Omega \times \mathbf{R}^m \to \mathbf{R}^*$, the weak limit is the function

$$\overline{\psi}(x) = \int_{\mathbf{R}^m} \psi(x,\lambda)\,d\nu_x(\lambda). \tag{6–5}$$

The idea of the proof is quite natural. The existence of ν is obtained in a simple way via the compactness property for weak * topologies in the appropriate space. The more technical part of the proof requires an extension of the representation of weak limits for Carathéodory functions.

Proof. We divide the proof in several steps.

Step 1. Existence of ν.

The vector space

$$\mathcal{C}_0(\mathbf{R}^m) = \left\{ f \in \mathcal{C}(\mathbf{R}^m) : \lim_{\lambda\to\infty} f(\lambda) = 0 \right\}$$

is a Banach space under the supremum norm. Its dual space is the space of Radon measures supported in $\mathbf{R}^m$ denoted $\mathcal{M}(\mathbf{R}^m)$ with the dual norm of the bounded variation. Since $\mathcal{C}_0(\mathbf{R}^m)$ is separable, we have, according to the discussion in the Appendix, Section 7, that

$$L^1(\Omega; \mathcal{C}_0(\mathbf{R}^m))' = L^\infty_{w*}(\Omega; \mathcal{M}(\mathbf{R}^m))$$

under the duality

$$\langle \psi, \mu \rangle = \int_\Omega \int_{\mathbf{R}^m} \psi(x, \lambda)\, d\mu_x(\lambda)\, dx,$$

for $\psi \in L^1(\Omega; \mathcal{C}_0(\mathbf{R}^m))$ and $\mu \in L^\infty_{w*}(\Omega; \mathcal{M}(\mathbf{R}^m))$. The norm in $L^\infty_{w*}(\Omega; \mathcal{M}(\mathbf{R}^m))$ is

$$\|\mu\| = \operatorname{ess\,sup}_{x\in\Omega} \|\mu_x\|_{\mathcal{M}(\mathbf{R}^m)}\,.$$

For each j, we define $\nu_j \in L^\infty_{w*}(\Omega; \mathcal{M}(\mathbf{R}^m))$ through the identification $\nu_j = \delta_{z_j(x)}$ where δ_a is the usual Dirac mass centered at $a \in \mathbf{R}^m$. For $\psi \in L^1(\Omega; \mathcal{C}_0(\mathbf{R}^m))$,

$$\begin{aligned}\langle \psi, \nu_j \rangle &= \int_\Omega \int_{\mathbf{R}^m} \psi(x, \lambda)\, d\delta_{z_j(x)}(\lambda)\, dx \\ &= \int_\Omega \psi(x, z_j(x))\, dx.\end{aligned}$$

It is easy to check that

$$\|\nu_j\| = 1, \quad \text{for every } j.$$

By the Banach-Alaouglu-Bourbaki theorem there exists some subsequence, not relabeled, and $\nu \in L^\infty_{w*}(\Omega; \mathcal{M}(\mathbf{R}^m))$ such that $\nu_j \overset{*}{\rightharpoonup} \nu$:

$$\lim_{j\to\infty} \int_\Omega \psi(x, z_j(x))\, dx = \int_\Omega \int_{\mathbf{R}^m} \psi(x, \lambda)\, d\nu_x(\lambda)\, dx, \tag{6–4}$$

for every $\psi \in L^1(\Omega; \mathcal{C}_0(\mathbf{R}^m))$.

Step 2. Some technical preliminaries.

Let ψ be a nonnegative, Carathéodory function such that $\{\psi(x, z_j(x))\}$ converges weakly in $L^1(\Omega)$. By Lemma 6.1,

$$\lim_{k\to\infty} \sup_j \int_{\{\psi(x,z_j(x))\geq k\}} \psi(x, z_j(x))\, dx = 0.$$

On the other hand since g is nondecreasing

$$g(k) \sup_j |\{|z_j| \geq k\}| \leq \sup_j \int_\Omega g(|z_j(x)|)\, dx < \infty,$$

and $\lim_{k\to\infty} g(k) = \infty$ implies that

$$\lim_{k\to\infty} \sup_j |\{|z_j| \geq k\}| = 0.$$

Therefore, we can choose $m_k \geq k$ in such a way that

$$\sup_j |\{|z_j| \geq m_k\}| \leq \frac{1}{k^2}.$$

Hence

$$k \sup_j |\{|z_j| \geq m_k\}| \to 0, \quad k \to \infty.$$

Finally, let θ^k be auxiliary functions defined for $t \in \mathbf{R}$ by

$$\theta^k(t) = \begin{cases} 1, & |t| \leq k, \\ 1 - |t| + k, & k \leq |t| \leq k+1, \\ 0, & |t| \geq k+1, \end{cases}$$

and $\psi^k(x, \lambda) = \theta^k(|\lambda|)\theta^k(\psi(x, \lambda))\psi(x, \lambda)$. It is then easy to deduce the following properties:

i) $\psi^k = \psi$ if $\psi \leq k$ and $|\lambda| \leq k$;
ii) $\psi^k \in L^1(\Omega; \mathcal{C}_0(\mathbf{R}^m))$ for all k;
iii) $0 \leq \psi^k \leq \psi$ for all k;
iv) $\{\psi^k\}$ is a non-decreasing sequence;
v) $\lim_{k\to\infty} \psi^k = \psi$ pointwise.

Step 3. Extension of (6–4).

In this step we would like to conclude that (6–4) is true under the assumptions in step 2. To this end, let

$$\gamma_{j,k} = \int_\Omega (\psi^{m_k}(x, z_j(x)) - \psi(x, z_j(x)))\, dx.$$

We have the following estimates

$$\begin{aligned} |\gamma_{j,k}| &\leq C \int_{\{|z_j| \geq m_k\} \cup \{\psi(x, z_j(x)) \geq m_k\}} \psi(x, z_j(x))\, dx \\ &\leq C \int_{\{|z_j| \geq m_k\} \cup \{\psi(x, z_j(x)) \geq k\}} \psi(x, z_j(x))\, dx \\ &\leq C \int_{\{\psi(x, z_j(x)) \geq k\}} \psi(x, z_j(x))\, dx \\ &\quad + C \int_{\{|z_j| \geq m_k\} \cap \{\psi(x, z_j(x)) \leq k\}} \psi(x, z_j(x))\, dx \\ &\leq C \sup_j \int_{\{\psi(x, z_j(x)) \geq k\}} \psi(x, z_j(x))\, dx \\ &\quad + Ck \sup_j |\{|z_j| \geq m_k\}|. \end{aligned}$$

By the discussion in step 2, we can conclude that

$$\lim_{k\to\infty} \int_\Omega \psi^{m_k}(x, z_j(x))\, dx = \int_\Omega \psi(x, z_j(x))\, dx$$

uniformly in j. In particular, this fact implies (elementary exercise) that

$$\lim_{j\to\infty}\lim_{k\to\infty}\int_\Omega \psi^{m_k}(x,z_j(x))\,dx = \lim_{k\to\infty}\lim_{j\to\infty}\int_\Omega \psi^{m_k}(x,z_j(x))\,dx.$$

Since $\psi^{m_k}\in L^1(\Omega;\mathcal{C}_0(\mathbf{R}^m))$ for all k, by (6–4),

$$\lim_{j\to\infty}\int_\Omega \psi(x,z_j(x))\,dx = \lim_{k\to\infty}\int_\Omega\int_{\mathbf{R}^m}\psi^{m_k}(x,\lambda)\,d\nu_x(\lambda)\,dx$$

and by the monotone convergence theorem in the second term (using iv) in step 2) we can conclude

$$\lim_{j\to\infty}\int_\Omega \psi(x,z_j(x))\,dx = \int_\Omega\int_{\mathbf{R}^m}\psi(x,\lambda)\,d\nu_x(\lambda)\,dx.$$

Step 4. Conclusion.

If we remove the nonnegativeness condition for ψ, we can always separate ψ in positive and negative parts, ψ^+ and ψ^- ($\psi^+ = \sup\{\psi,0\}$, $\psi^- = \sup\{-\psi,0\}$) and apply steps 2 and 3 to these two functions, bearing in mind that the weak convergence in $L^1(\Omega)$ brings along the equiintegrability of the sequence $\{|\psi(x,z_j(x))|\}$ and therefore the equiintegrability for ψ^+ and ψ^-. Notice that $\psi=\psi^+-\psi^-$ and $|\psi|=\psi^++\psi^-$.

For $\xi\in L^\infty(\Omega)$ we can take $\varphi(x,\lambda)=\xi(x)\psi(x,\lambda)$, so that φ is a Carathéodory function itself to which we can apply the preceding arguments. Observe that the weak convergence in $L^1(\Omega)$ of the sequence $\{\psi(x,z_j(x))\}$ implies the same for $\{\xi(x)\psi(x,z_j(x))\}$. Thus (6–4) also holds for φ, and since $\xi\in L^\infty(\Omega)$ is arbitrary, we obtain

$$\psi(x,z_j(x)) \rightharpoonup \overline{\psi}(x) = \int_{\mathbf{R}^m}\psi(x,\lambda)\,d\nu_x(\lambda)\,dx$$

in $L^1(\Omega)$.

Finally, it is not hard to check that almost every ν_x is a probability measure. By weak lower semicontinuity of the norm

$$\|\nu\| \le \liminf_{j\to\infty}\left\|\delta_{z_j}\right\| = 1,$$

so that $\|\nu_x\|_{\mathcal{M}(\mathbf{R}^m)}\le 1$ for a.e. $x\in\Omega$. If we take in particular $\psi=\chi_{B_R}(x)$ for B_R the ball of radius R centered at the origin in (6–4), then

$$\int_{B_R\cap\Omega}\int_{\mathbf{R}^m} 1\,d\nu_x(\lambda)\,dx = \lim_{j\to\infty}\int_{B_R\cap\Omega}1\,dx = |B_R\cap\Omega|\,.$$

Therefore

$$\begin{aligned}|B_R\cap\Omega| &= \int_{B_R\cap\Omega}\int_{\mathbf{R}^m}1\,d\nu_x(\lambda)\,dx\\ &\le \int_{B_R\cap\Omega}\|\nu_x\|\,dx\\ &\le |B_R\cap\Omega|,\end{aligned}$$

and ν_x is equal to its total variation for a.e. $x\in\Omega$, i.e., $\nu_x\ge 0$ and $\|\nu_x\|_{\mathcal{M}(\mathbf{R}^m)}=1$. ∎

A particularly important example is obtained by taking $g(t) = t^p$ for $p \geq 1$ (we can also allow $0 < p < 1$). In this case, *every bounded sequence in $L^p(\Omega)$ contains a subsequence that generates a parametrized measure* in the sense of Theorem 6.2.

An important remark to bear in mind when working with parametrized measures is that in order to identify the parametrized measure associated to a particular sequence of functions $\{z_j\}$ (obtained perhaps in some constructive way or using some scheme), it is enough to check

$$\varphi(z_j) \overset{*}{\rightharpoonup} \overline{\varphi} \quad \text{in } L^\infty(\Omega),$$

for every $\varphi \in \mathcal{C}_0(\mathbf{R}^m)$ where as usual

$$\overline{\varphi}(x) = \int_{\mathbf{R}^m} \varphi(\lambda)\, d\nu_x(\lambda).$$

It is even enough to have

$$\lim_{j\to\infty} \int_\Omega \xi(x)\varphi(z_j(x))\, dx = \int_\Omega \xi(x) \int_{\mathbf{R}^m} \varphi(\lambda)\, d\nu_x(\lambda)\, dx \tag{6–5}$$

for ξ and φ belonging to dense, countable subsets of $L^1(\Omega)$ and $\mathcal{C}_0(\mathbf{R}^m)$, respectively. If this is so for a given family of probability measures $\nu = \{\nu_x\}_{x\in\Omega}$ and $\{z_j\}$ satisfies (6–2) then ν is the parametrized measure associated to $\{z_j\}$ and therefore

$$\psi(x, z_j(x)) \rightharpoonup \overline{\psi}(x) = \int_{\mathbf{R}^m} \psi(x, \lambda)\, d\nu_x(\lambda),$$

for every Carathéodory function ψ such that $\{\psi(x, z_j(x))\}$ is weakly convergent in $L^1(\Omega)$. The reason for this is that probability measures ν are identified by their action on $\mathcal{C}_0(\mathbf{R}^m)$. (6–5) identifies each ν_x for a.e. $x \in \Omega$.

There are two interesting situations where this remark will have some relevance for us. For reference, we include them in the following lemma.

Lemma 6.3 *Assume that we have two sequences, $\{z_j\}$ and $\{w_j\}$, both bounded in $L^p(\Omega)$.*

i) If $|\{z_j \neq w_j\}| \to 0$, the parametrized measure for both sequences is the same.

ii) If

$$\|z_j - w_j\|_{L^p(\Omega)} \to 0,$$

$\{z_j\}$ and $\{w_j\}$ share the parametrized measure.

Proof. The proof is simple. Let $\varphi \in \mathcal{C}_0(\mathbf{R}^m)$ and $\xi \in L^1(\Omega)$. Then

$$\left| \int_\Omega \xi(x)\varphi(z_j(x))\,dx - \int_\Omega \xi(x)\varphi(w_j(x))\,dx \right| \leq \int_{\{z_j \neq w_j\}} 2\,\|\varphi\|_{L^\infty(\mathbf{R}^m)}\,|\xi(x)|\,dx.$$

The integrand on the right-hand side is a $L^1(\Omega)$-function and it is integrated over a sequence of sets of vanishing measure. Hence the limit vanishes as $j \to \infty$, and this in turn implies that the weak limits for $\{\varphi(z_j)\}$ and $\{\varphi(w_j))\}$ are the same. By the above remark both sequences share the parametrized measure.

For ii), use the dominated convergence theorem to examine the difference

$$\int_\Omega \xi(x)\left[\varphi(z_j(x)) - \varphi(w_j(x))\right] dx$$

for $\xi \in L^1(\Omega)$ and $\varphi \in \mathcal{C}_0(\mathbf{R}^m)$. ∎

A helpful example of this situation is the following. Assume $\{z_j\}$ is uniformly bounded in $L^p(\Omega)$ and let $\nu = \{\nu_x\}_{x\in\Omega}$ be its associated parametrized measure. Consider the truncation operators

$$T_k(\lambda) = \begin{cases} \lambda, & |\lambda| \leq k, \\ k\dfrac{\lambda}{|\lambda|}, & |\lambda| > k. \end{cases}$$

We claim that for any subsequence $k(j) \to \infty$ as $j \to \infty$ the parametrized measure corresponding to $\left\{T_{k(j)}(z_j)\right\}$ is also ν. To this aim, we simply notice that

$$|\{|z_j| > k(j)\}| \leq \frac{\sup_j \|z_j\|_{L^p(\Omega)}}{k(j)^p} \to 0$$

if $k(j) \to \infty$.

3. Sufficient conditions for representation of weak limits

A crucial point in Theorem 6.2 in order to have the representation (6–5) of weak limits in terms of integrals against the parametrized measure is to have "a priori" the weak convergence in $L^1(\Omega)$ of the sequence whose weak limit we are concerned about. We have also insisted upon the fact that uniform bounds in $L^1(\Omega)$ are not sufficient to ensure this weak convergence. Equiintegrability should be taken care of. For this reason it is important to rely on criteria that under suitable hypotheses enable us to ascertain this equiintegrability.

One such important weak compactness criteron in $L^1(\Omega)$ that can be used in some situations is the De La Vallé-Poussin criterion. The sufficiency of such a result is an immediate consequence of Lemma 6.1, and this sufficiency is in fact the part of the result which is most useful.

Lemma 6.4 *(De La Vallé-Poussin criterion) Let Ω be bounded. The sequence $\{f_j\}$ is sequentially weakly relatively compact in $L^1(\Omega)$ if and only if*

$$\sup_j \int_\Omega \psi\left(|f_j|\right)\, dx < \infty, \tag{6–6}$$

for some continuous function $\psi : [0, \infty) \to \mathbf{R}$ with

$$\lim_{\lambda \to \infty} \frac{\psi(\lambda)}{\lambda} = \infty. \tag{6–7}$$

For the sufficiency, let us suppose that there is a function ψ satisfying (6–6) and (6–7). We want to show that (6–1) is true. For $\epsilon > 0$, take M such that $M\epsilon \geq C$ where

$$C = \sup_j \int_\Omega \psi(|f_j|)\, dx < \infty.$$

Because of (6–7), and for k large enough, $M\lambda \leq \psi(\lambda)$ if $\lambda \geq k$. Thus

$$\int_{\{|f_j| \geq k\}} |f_j|\; dx \leq \frac{1}{M} \int_{\{|f_j| \geq k\}} \psi(|f_j|)\, dx \leq \frac{1}{M} C \leq \epsilon,$$

uniformly in j. Therefore the sequence is weakly relatively compact by Lemma 6.1.

In the spirit of this criterion we can give the following proposition that allows us to have the representation of weak limits in terms of parametrized measures. According to Theorem 6.2 this requires the weak convergence in $L^1(\Omega)$ of the sequence involved.

Proposition 6.5 *Let $\psi : \Omega \times \mathbf{R}^m \to \mathbf{R}$ be a Carathéodory function such that*

$$|\psi(x, \lambda)| \leq \tilde{\psi}(|\lambda|), \text{ for a.e. } x \in \Omega,$$

where $\tilde{\psi} \in L^\infty_{loc}(\mathbf{R})$. Let $\{z_j\}$ be a sequence with

$$\sup_j \int_\Omega g(|z_j|)\, dx = C < \infty,$$

where g is a continuous, nondecreasing function and $\lim_{t\to\infty} g(t) = \infty$. If

$$\lim_{t \to \infty} \frac{\tilde{\psi}(t)}{g(t)} = 0 \tag{6–8}$$

then

$$\psi(x, z_j(x)) \rightharpoonup \overline{\psi}(x) = \int_{\mathbf{R}^m} \psi(x, \lambda)\, d\nu_x(\lambda) \quad \text{in } L^1(\Omega),$$

where $\nu = \{\nu_x\}_{x\in\Omega}$ is the parametrized measure associated to the sequence $\{z_j\}$ (or possibly to a subsequence).

Proof. Notice that the conclusion of the proposition is that

$$\lim_{j\to\infty}\int_{\Omega}\xi(x)\psi(x,z_j(x))\,dx=\int_{\Omega}\xi(x)\int_{\mathbf{R}^m}\psi(x,\lambda)\,d\nu_x(\lambda)\,dx \tag{6–9}$$

for all $\xi\in L^\infty(\Omega)$.

Since $\tilde{\psi}\in L^\infty_{loc}(\mathbf{R})$, choose $m_k\to\infty$ such that

$$\left\{\tilde{\psi}(|\lambda|)\geq k\right\}\subset\{|\lambda|\geq m_k\}.$$

Then

$$\begin{aligned}\int_{\{|\psi(x,z_j(x))|\geq k\}}|\psi(x,z_j(x))|\,dx&\leq\int_{\left\{\tilde{\psi}(|z_j|)\geq k\right\}}\tilde{\psi}(|z_j|)\,dx\\&\leq\int_{\{|z_j|\geq m_k\}}\tilde{\psi}(|z_j|)\,dx\\&\leq\frac{1}{M_k}\int_{\Omega}g(|z_j|)\,dx\\&\leq\frac{C}{M_k}\to 0,\end{aligned}$$

uniformly in j, where $g(t)\geq M_k\tilde{\psi}(t)$ for $t\geq m_k$ and $M_k\to\infty$ by (6–8).

This implies the weak convergence in $L^1(\Omega)$ of $\{\psi(x,z_j(x))\}$ and thus the representation (6–9) holds. ∎

A particular, important example is $g(t)=t^p$, $p>0$ and $\tilde{\psi}(t)=t^q$, $p>q>0$. In this case we have the representation (6–9) when the sequence $\{z_j\}$ is uniformly bounded in $L^p(\Omega)$ and $|\psi(x,\lambda)|\leq|\lambda|^q$. However, Proposition 6.5 fails if $p=q$, so that for functions ψ that grow like the pth power in λ the representation (6–9) may not be valid. This brings us to the question of what is the relationship between both terms in (6–9) in this situation when we do not have equality. In order to understand this question it is convenient to introduce the notion of biting convergence and compare it to weak convergence. We are going to explore this issue in subsequent sections.

We close this section with a remarkable example. When equiintegrability fails, concentrations may develop even in a rather nasty way. This phenomenon is responsible for failure of the representation (6–5). Our example is one-dimensional. Consider the sequence of functions defined on $\Omega=(0,1)$ by

$$f_j(x)=\begin{cases}j^2/2, & \text{for } x\in(k(j+1)^{-1}-j^{-3},k(j+1)^{-1}+j^{-3}),\ k=1,2,\dots,j,\\ 0, & \text{otherwise.}\end{cases}$$

Then $\|f_j\|_{L^1(\Omega)} = 1$ for all j, and for φ continuous

$$\begin{aligned}\int_0^1 \varphi(x) f_j(x)\,dx &= \sum_{k=1}^{j} \int_{k(j+1)^{-1}-j^{-3}}^{k(j+1)^{-1}+j^{-3}} \frac{j^2}{2}\varphi(x)\,dx \\ &= \frac{j^2}{2}\sum_{k=1}^{j} \frac{2}{j^3}\varphi(x_k) \\ &= \frac{1}{j}\sum_{k=1}^{j}\varphi(x_k) \\ &\to \int_0^1 \varphi(x)\,dx,\end{aligned}$$

where the points $x_k \in (k(j+1)^{-1} - j^{-3}, k(j+1)^{-1} + j^{-3})$. Hence the sequence $\{f_j\}$ converges weak * in the sense of measures to 1. For r fixed, if $j^2/2 \geq r$ then $\{|f_j| \geq r\} = \{f_j \neq 0\}$ and

$$\int_{\{|f_j|\geq r\}} |f_j|\,dx = \frac{j^2}{2}\frac{2}{j^2} = 1.$$

Therefore

$$\lim_{r\to\infty} \sup_j \int_{\{|f_j|\geq r\}} |f_j|\,dx \geq 1,$$

and by Lemma 6.1 the sequence cannot be weak convergent in $L^1(\Omega)$. What is the parametrized measure associated to $\{f_j\}$? We will answer this question after the discussion of the next section. Note how this example also illustrates that convergence in the sense of distributions and pointwise convergence are different.

4. Chacon's biting lemma and biting convergence

Whenever a bounded sequence in $L^1(\Omega)$ is not equiintegrable, one can "remove" the set where concentrations occur and be left with a well-behaved sequence. This is essentially what Chacon's biting lemma says. The proof can be done in a very general and abstract setting. We restrict attention, however, to the framework in which we will be using this fact.

Theorem 6.6 *(Chacon's biting lemma) Let $\{f_j\}$ be a uniformly bounded sequence in $L^1(\Omega)$,*

$$\sup_j \|f_j\|_{L^1(\Omega)} = C < \infty.$$

There exists a subsequence, not relabeled, a nonincreasing sequence of measurable sets $\Omega_n \subset \Omega$, $|\Omega_n| \searrow 0$ and $f \in L^1(\Omega)$ such that

$$f_j \rightharpoonup f \text{ in } L^1(\Omega \setminus \Omega_n)$$

for all n.

Proof. For $j, k \in \mathbf{N}$ set

$$\alpha_{j,k} = \int_{\{|f_j| \geq k\}} |f_j| \, dx \geq 0.$$

Notice that the sequence $\{\sup_j \alpha_{j,k}\}$ is monotone and nonincreasing. Let

$$L = \lim_{k\to\infty} \sup_j \alpha_{j,k} \geq 0.$$

If $L = 0$, by Lemma 6.1 we can take $\Omega_n = \emptyset$ for all n because in this case weak convergence in $L^1(\Omega)$ holds for some subsequence.

Let us just assume that $L > 0$. For each $m \in \mathbf{N}$, let j_m be such that

$$\alpha_{j_m, 2^m} \geq \sup_j \alpha_{j,2^m} - \frac{1}{m}.$$

In this way,

$$\alpha_{j_m, 2^m} \geq L - \frac{1}{m}. \tag{6–10}$$

By monotonicity there also exists the limit

$$\lim_{r\to\infty} \sup_m \int_{\{r \leq |f_{j_m}| < 2^m\}} |f_{j_m}| \, dx = L' \geq 0,$$

and in particular there exists a subsequence $m(r)$ for r sufficiently large, such that

$$\int_{\left\{r \leq \left|f_{j_{m(r)}}\right| < 2^{m(r)}\right\}} \left|f_{j_{m(r)}}\right| \, dx \geq \frac{L'}{2}.$$

Finally, bearing in mind the monotonicity of $\{\sup_j \alpha_{j,k}\}$,

$$\begin{aligned}
L &= \lim_{r\to\infty} \sup_j \alpha_{j,r} \geq \lim_{r\to\infty} \sup_m \alpha_{j_m,r} \geq \lim_{r\to\infty} \alpha_{j_{m(r)},r} \\
&= \lim_{r\to\infty} \left(\int_{\left\{r \leq \left|f_{j_{m(r)}}\right| < 2^{m(r)}\right\}} \left|f_{j_{m(r)}}\right| \, dx + \int_{\left\{\left|f_{j_{m(r)}}\right| \geq 2^{m(r)}\right\}} \left|f_{j_{m(r)}}\right| \, dx \right) \\
&\geq \frac{L'}{2} + \lim_{r\to\infty} \alpha_{j_{m(r)}, 2^{m(r)}} \geq \frac{L'}{2} + L.
\end{aligned}$$

The last inequality holds because of (6–10). Hence $L' = 0$, i.e.,

$$\lim_{r\to\infty} \left(\sup_m \int_{\{r \leq |f_{j_m}| < 2^m\}} |f_{j_m}| \, dx \right) = 0. \tag{6–11}$$

Define the exceptional set Ω_n by

$$\Omega_n = \bigcup_{m=n}^{\infty} \{|f_{j_m}| \geq 2^m\} \subset \Omega.$$

We have the claimed properties:

i) $\Omega_{n+1} \subset \Omega_n$, for all n;

ii) $|\Omega_n| \searrow 0$:

$$2^m \, |\{|f_{j_m}| \geq 2^m\}| \leq \int_{\{|f_{j_m}| \geq 2^m\}} |f_{j_m}| \, dx \leq C$$

and,

$$|\Omega_n| \leq \sum_{m=n}^{\infty} |\{|f_{j_m}| \geq 2^m\}| \leq C \sum_{m=n}^{\infty} \frac{1}{2^m}.$$

This last term tends to 0 as n tends to ∞.

iii) For fixed n and keeping in mind (6–11),

$$\begin{aligned} 0 &\leq \lim_{r\to\infty} \sup_{m \geq n} \int_{r \leq \{|f_{j_m}|\} \setminus \Omega_n} |f_{j_m}| \, dx \\ &\leq \lim_{r\to\infty} \sup_{m \geq n} \int_{\{r \leq |f_{j_m}| < 2^m\}} |f_{j_m}| \, dx \leq 0. \end{aligned}$$

This implies, by Lemma 6.1, that the sequence $\{f_{j_m}\}$ is sequentially weakly relatively compact in $L^1(\Omega \setminus \Omega_n)$ for all n. The following technical proposition, whose proof is left as an exercise to the interested reader, helps us in concluding the proof.

Proposition 6.7 *Let $\{f_j\}$ be a sequence of functions bounded in $L^1(\Omega)$ such that there exist sets $\Omega_n \subset \Omega$, $\Omega_{n+1} \subset \Omega_n$, $|\Omega_n| \searrow 0$ with $\{f_j\}$ sequentially weakly relatively compact in $L^1(\Omega \setminus \Omega_n)$ for all n. There exists a subsequence $\{f_{j_k}\}$ and $f \in L^1(\Omega)$ such that*

$$f_{j_k} \rightharpoonup f \text{ in } L^1(\Omega \setminus \Omega_n)$$

for all n. ■

In the example given at the end of the previous section, the exceptional sets are precisely $\{f_j \neq 0\}$ and then trivially $f_j \rightharpoonup 0$ in $L^1(\Omega \setminus \{f_k \neq 0\})$ for all k.

Even when the weak convergence in $L^1(\Omega)$ fails, we can still consider the function $\overline{\psi}$ as defined in (6–5). This function can no longer be the weak limit of $\{\psi(x, z_j(x))\}$ as we have placed ourselves precisely in the situation when there is no weak limit at all. Yet, one might ask what is the relationship between $\overline{\psi}$ and $\{\psi(x, z_j(x))\}$ if any. We are led to introduce biting convergence.

The sequence $\{f_j\} \subset L^1(\Omega)$ converges in the biting sense to $f \in L^1(\Omega)$ and is denoted

$$f_j \stackrel{b}{\rightharpoonup} f \text{ in } L^1(\Omega),$$

if there exists a nonincreasing sequence of measurable sets $\{\Omega_n\}$ such that $|\Omega_n| \searrow 0$ and

$$f_j \rightharpoonup f \text{ in } L^1(\Omega \setminus \Omega_n)$$

for all n. We may restate Chacon's biting lemma by saying that a uniformly bounded sequence in $L^1(\Omega)$ contains a subsequence converging in the biting sense to a function in $L^1(\Omega)$.

The relationship between biting convergence and parametrized measures is given in the following theorem.

Theorem 6.8 *Let $\{z_j\}$ be a sequence of vector-valued functions with associated parametrized measure $\nu = \{\nu_x\}_{x\in\Omega}$. If $\varphi : \Omega \times \mathbf{R}^m \to \mathbf{R}^*$ is a Carathéodory function such that the sequence $\{\varphi(x, z_j(x))\}$ is uniformly bounded in $L^1(\Omega)$, then possibly for a subsequence*

$$\varphi(x, z_j(x)) \stackrel{b}{\rightharpoonup} \overline{\varphi}(x) = \int_{\mathbf{R}^m} \varphi(x, \lambda)\, d\nu_x(\lambda)\, dx. \tag{6–12}$$

Proof. The proof is elementary. By Chacon's biting lemma, there exists a collection of subsets $\{\Omega_n\}$ and $\tilde{\varphi} \in L^1(\Omega)$ such that $|\Omega_n| \searrow 0$ and

$$\varphi(x, z_j(x)) \rightharpoonup \tilde{\varphi} \text{ in } L^1(\Omega \setminus \Omega_n)$$

for all n. By Theorem 6.2 whenever weak convergence in $L^1(E)$ holds for any subset $E \subset \Omega$, the weak limit has to be $\overline{\varphi}$ in (6–12). Since $|\Omega_n| \searrow 0$, we conclude that $\tilde{\varphi} = \overline{\varphi}$ a.e. $x \in \Omega$. ∎

Parametrized measures always yield biting limits under the very mild assumption of boundedness in $L^1(\Omega)$. Having this remark in mind, it is easy to conclude that the parametrized measure corresponding to the one-dimensional example described in the previous section is δ_0, the Dirac mass centered at 0, for all $x \in (0, 1)$.

5. Biting convergence and weak convergence

In some circumstances biting convergence may be improved to weak convergence so that parametrized measures will provide weak limits. This amounts to discarding the possibility of concentrations. The following lemma gives a necessary and sufficient condition for such an improvement: biting convergence may become in fact weak convergence.

Lemma 6.9 *Let $f_j : \Omega \to \mathbf{R}^+$ ($f_j \geq 0$) be a sequence of measurable functions in $L^1(\Omega)$, converging in the biting sense to $f \in L^1(\Omega)$.*

A subsequence converges weakly in $L^1(\Omega)$ if and only if

$$\liminf_{j\to\infty} \int_\Omega f_j(x)\,dx \leq \int_\Omega f(x)\,dx. \tag{6–13}$$

Moreover, the whole sequence $\{f_j\}$ converges weakly in $L^1(\Omega)$ to f if and only if

$$\limsup_{j\to\infty} \int_\Omega f_j(x)\,dx \leq \int_\Omega f(x)\,dx. \tag{6–14}$$

Proof. The proof is simple. Assuming biting convergence and the failure of the Dunford-Pettis criterion leads to the failure of (6–13).

Let $\{\Omega_n\} \subset \Omega$ be the sequence of subsets associated with the biting convergence so that $|\Omega_n| \searrow 0$, $\Omega_{n+1} \subset \Omega_n$ and

$$f_j \rightharpoonup f \text{ in } L^1(\Omega \setminus \Omega_n)$$

for all n. We may assume, working with subsequences, that

$$\lim_{j\to\infty} \int_\Omega f_j\,dx \leq \int_\Omega f\,dx < \infty. \tag{6–15}$$

Suppose that no subsequence converges weakly in $L^1(\Omega)$. By the Dunford-Pettis criterion, there exists $\epsilon > 0$ and a subsequence $\{k_n\}$ such that

$$\epsilon \leq \int_{\Omega_n} f_{k_n}\,dx$$

for all n sufficiently large, because outside Ω_n we do have weak convergence. In particular, if $i > n$, since $f_{k_i} \geq 0$,

$$\epsilon \leq \int_{\Omega_i} f_{k_i}\,dx \leq \int_{\Omega_n} f_{k_i}\,dx,$$

and for fixed n, $i > n$,

$$\begin{aligned}\int_\Omega f_{k_i}\,dx &= \int_{\Omega_n} f_{k_i}\,dx + \int_{\Omega\setminus\Omega_n} f_{k_i}\,dx \\ &\geq \epsilon + \int_{\Omega\setminus\Omega_n} f_{k_i}\,dx.\end{aligned}$$

Finally, letting $i \to \infty$,

$$\lim_{i\to\infty} \int_\Omega f_{k_i}\,dx \geq \epsilon + \int_{\Omega\setminus\Omega_n} f\,dx.$$

This is true for every n, and consequently

$$\lim_{i\to\infty} \int_\Omega f_{k_i}\,dx \geq \epsilon + \int_\Omega f\,dx,$$

contrary to (6–15). ∎

A straightforward corollary is the following fact whose proof is left as an exercise.

Corollary 6.10 *Let $\{z_j\}$ be a sequence of vector valued functions with associated parametrized measure $\nu = \{\nu_x\}_{x\in\Omega}$. If for φ_0, a nonnegative Carathéodory function, we have*

$$\lim_{j\to\infty} \int_\Omega \varphi_0(x, z_j(x))\,dx = \int_\Omega \int_{\mathbf{R}^m} \varphi_0(x,\lambda)\,d\nu_x(\lambda)\,dx < \infty,$$

then

$$\lim_{j\to\infty} \int_E \varphi(x, z_j(x))\,dx = \int_E \int_{\mathbf{R}^m} \varphi(x,\lambda)\,d\nu_x(\lambda)\,dx < \infty,$$

for any measurable subset $E \subset \Omega$ and for any φ in the space

$$\mathcal{E}^{\varphi_0} = \{\varphi,\ \text{Carathéodory functions}, |\varphi| \le C(1+\varphi_0)\}.$$

If in spite of all efforts Corollary 6.10 cannot be applied so that concentrations may arise, we still can draw some information that might be helpful in some circumstances.

Theorem 6.11 *If $\{z_j\}$ is a sequence of measurable functions with associated parametrized measure $\nu = \{\nu_x\}_{x\in\Omega}$,*

$$\liminf_{j\to\infty} \int_E \psi(x, z_j(x))\,dx \ge \int_E \int_{\mathbf{R}^m} \psi(x,\lambda)\,d\nu_x(\lambda)\,dx, \tag{6–16}$$

for every nonnegative, Carathéodory function ψ and every measurable subset $E \subset \Omega$.

Proof. If the left-hand side of (6–16) is infinite, there is nothing to be proved. If it is finite, the sequence $\{\psi(x, z_j(x))\}$ is a bounded sequence in $L^1(E)$. If we set as usual

$$\overline{\psi}(x) = \int_{\mathbf{R}^m} \psi(x,\lambda)\,d\nu_x(\lambda),$$

then

$$\psi(x, z_j(x)) \overset{b}{\rightharpoonup} \overline{\psi} \quad \text{in } L^1(E).$$

By Lemma 6.9, it is not possible to have the strict inequality

$$\int_E \overline{\psi}(x)\,dx > \liminf_{j\to\infty} \int_E \psi(x, z_j(x))\,dx.$$ ∎

Strict inequality in (6–16) occurs when the sequence $\{\psi(x, z_j(x))\}$ develops concentrations. In this sense we say that parametrized measures do not capture concentration effects.

It is obvious that Theorem 6.11 still holds true if ψ is bounded from below by some constant.

6. Strong convergence

We would like to understand how strong convergence gets translated into the parametrized measure. A first thought is that since parametrized measures are a device to keep track of oscillations, and strong convergence rules out this phenomenon, one can expect that parametrized measures associated with strong convergent sequences are trivial. In this section we restrict attention to the case in which $g(t) = t^p$.

Proposition 6.12 *Let $\{z_j\}$ be a sequence in $L^p(\Omega)$ such that $\{|z_j|^p\}$ is weakly convergent in $L^1(\Omega)$ for $p < \infty$ and $\nu = \{\nu_x\}_{x\in\Omega}$ is the associated parametrized measure. $z_j \to z$ strongly in $L^p(\Omega)$ if and only if $\nu_x = \delta_{z(x)}$ for a.e. $x \in \Omega$.*

Proof. Let us consider the Carathéodory function $\psi(x, \lambda) = |\lambda - z(x)|^p$. Because of the hypothesis on $\{z_j\}$ when $p < \infty$, the sequence $\{\psi(x, z_j(x))\}$ is weakly convergent in $L^1(\Omega)$ and therefore the integral representation in terms of ν is correct

$$\lim_{j\to\infty} \int_\Omega \psi(x, z_j(x))\, dx = \int_\Omega \int_{\mathbf{R}^m} \psi(x, \lambda)\, d\delta_{z(x)}(\lambda)\, dx = 0,$$

whence $z_j \to z$ strong in $L^p(\Omega)$.

Conversely, if $z_j \to z$ strong in $L^p(\Omega)$, for any continuous, bounded function $\psi(\lambda)$, we would have $\psi(z_j) \to \psi(z)$ strong in $L^p(\Omega)$. This implies, in particular, that for any measurable $E \subset \Omega$,

$$\int_E \psi(z(x))\, dx = \int_E \int_{\mathbf{R}^m} \psi(\lambda)\, d\nu_x(\lambda)\, dx.$$

We can conclude that

$$\psi(z(x)) = \int_{\mathbf{R}^m} \psi(\lambda)\, d\nu_x(\lambda)$$

for a.e. $x \in \Omega$.

The arbitrariness of ψ leads to $\nu_x = \delta_{z(x)}$ for a.e. $x \in \Omega$. ■

The condition on the weak convergence of $\{|z_j|^p\}$ for $p < \infty$ is necessary as the one-dimensional example studied in Section 3 shows. Notice also that this fact is not true for $p = \infty$. Take $\Omega = (0, 1)$ and $z_j = x^j$ (jth powers) for $x \in (0, 1)$. It is easy to find that $\nu = \delta_0$ but $\{z_j\}$ does not converge strongly to 0 in $L^\infty(\Omega)$. What at least is true is the fact that being the parametrized measure a delta prevents oscillations.

It is also helpful to consider parametrized measures coming from sequences for which we have strong convergence only for some components of the sequence but not for all of them. In this case strong convergence reflects triviality of the parametrized measure for the corresponding components.

Proposition 6.13 *Let $z_j = (u_j, v_j) : \Omega \to \mathbf{R}^d \times \mathbf{R}^m$ be a bounded sequence in $L^p(\Omega)$ such that $\{u_j\}$ converges strongly to u in $L^p(\Omega)$. If $\nu = \{\nu_x\}_{x\in\Omega}$ is the parametrized measure associated with $\{z_j\}$, $\nu_x = \delta_{u(x)} \otimes \mu_x$ a.e. $x \in \Omega$, where $\{\mu_x\}_{x\in\Omega}$ is the parametrized measure corresponding to $\{v_j\}$.*

Proof. Let $\psi_1 : \mathbf{R}^d \to \mathbf{R}$ and $\psi_2 : \mathbf{R}^m \to \mathbf{R}$ be continuous, bounded functions, so that

$$\psi_1(u_j) \to \psi_1(u) \text{ in } L^p(\Omega),$$
$$\psi_2(v_j) \rightharpoonup \overline{\psi}_2(x) = \int_{\mathbf{R}^m} \psi_2(\lambda)\, d\mu_x(\lambda) \text{ in } L^q(\Omega), \quad \frac{1}{p} + \frac{1}{q} = 1.$$

(In fact, $\psi_2(v_j) \overset{*}{\rightharpoonup} \overline{\psi}_2(x)$ in $L^\infty(\Omega)$ if ψ_2 is bounded.) In this case,

$$\psi_1(u_j)\psi_2(v_j) \rightharpoonup \psi_1(u)\overline{\psi}_2(x) \quad \text{in } L^1(E)$$

for any $E \subset \Omega$ (this is easy to check) and therefore

$$\int_E \int_{\mathbf{R}^d\times\mathbf{R}^m} \psi_1(\lambda_1)\psi_2(\lambda_2)\, d\nu_x(\lambda_1, \lambda_2)\, dx$$
$$= \int_E \int_{\mathbf{R}^d\times\mathbf{R}^m} \psi_1(\lambda_1)\psi_2(\lambda_2)\, d(\delta_{u(x)}(\lambda_1) \otimes \mu_x(\lambda_2))\, dx.$$

The arbitrariness of ψ_1, ψ_2 and E proves the result. ■

We have already see the relevance of this proposition in dealing with variational principles (Chapters 2 and 3).

7. Appendix

We need to give a few basic notions of L^p-spaces when the target space for functions is some general Banach space X with dual X'. For $\Omega \subset \mathbf{R}^N$ we write

$$L^p(\Omega; X) = \{f : \Omega \to X : f \text{ is strongly measurable and } \int_\Omega \|f(x)\|_X^p\, dx < \infty\}.$$

Such a function f is said to be strongly measurable if there exists a sequence of simple, measurable functions $\{f_j\}$ such that $f_j(x) \to f(x)$ a.e. $x \in \Omega$ and

$$\int_\Omega \|f_j(x) - f_k(x)\|_X^p\, dx \to 0, \quad j, k \to \infty.$$

We write

$$L^p_w(\Omega; X) = \{f : \Omega \to X : f \text{ is weakly measurable, } \|f(x)\|_X \text{ is a measurable function of } x \text{ and } \int_\Omega \|f(x)\|_X^p\, dx < \infty\}.$$

A function f is weakly measurable if for every $T \in X'$ the function of x, $x \mapsto \langle f(x), T\rangle$ is measurable. In the same way

$$L^p_{w*}(\Omega; X') = \{f : \Omega \to X' : f \text{ is weakly * measurable, } \|f(x)\|_{X'} \text{ is a measurable function of } x \text{ and } \int_\Omega \|f(x)\|^p_{X'}\, dx < \infty\Big\}.$$

$L^p(\Omega; X)$, $L^p_w(\Omega; X)$ and $L^p_{w*}(\Omega; X')$ are Banach spaces under the L^p-norm.

Theorem 6.14 *Let X be a separable Banach space with dual X'. Then*

$$(L^p(\Omega; X))' = L^q_{w*}(\Omega; X'), \quad 1 \le p < \infty, \frac{1}{p} + \frac{1}{q} = 1,$$

under the duality

$$\langle f, g\rangle = \int_\Omega \langle f(x), g(x)\rangle\, dx$$

where $f \in L^p(\Omega; X)$ and $g \in L^q_{w*}(\Omega; X')$.

The particular case we are interested in is

$$X = \mathcal{C}_0(\mathbf{R}^m) = \Big\{f : \mathbf{R}^m \to \mathbf{R} : \lim_{\lambda\to\infty} f(\lambda) = 0\Big\},$$
$$X' = \mathcal{M}(\mathbf{R}^m) = \{\text{ bounded, Radon measures on } \mathbf{R}^m\}.$$

In this case we have the duality

$$(L^1(\Omega; \mathcal{C}_0(\mathbf{R}^m)))' = L^\infty_{w*}(\Omega; \mathcal{M}(\mathbf{R}^m)).$$

8. Bibliographical remarks

General treatments of parametrized measures and applications to different problems can be found in [24], [27], [46], [212], [213]. It is interesting to have a look at the original papers of Young, [314] and [315]. See also [93] and [302]. The existence theorem presented here is basically the version contained in [31], with some variants. In [132], there is an existence theorem based on the technique of slicing measures.

With regard to general books on measure theory, we have included several choices: [47], [164], [214], [283]. See also [136]. A complete analysis of the De La Vallée-Poussin criterion for weak convergence in L^1 may be found in [214]. The one-dimensional example included in Section 3 has been taken from [39], in connection with Chacon's biting lemma. This is a good reference for a more general proof of Chacon's lemma. See also [58]. The proof that has been carried out here for this lemma is a less ambitious, more direct one. The book [116] is a good source for an accesible treatment of L^p-spaces when the target space is another Banach space. In particular the duality Theorem 6.14 has been taken from this reference.

Biting convergence has also been considered in a different setting in [317]. Some of the ideas explored in this chapter have been also studied in [40], [191] and [260].

Chapter 7
Analysis of Parametrized Measures

1. Introduction

In this chapter we shall analyze more closely parametrized measures and introduce the basic tools to deal with these families of probability measures. Some of these will be used several times later. Our main goal here is to characterize parametrized measures: we are interested in knowing when a given family of probability measures can actually be generated as the parametrized measure by some sequence of functions. At this stage we do not place any further restriction on the sequences we would like to consider except for boundedness in some $L^p(\Omega)$. In this regard we place ourselves in the context of Section 2 of Chapter 2. As a matter of fact, the main theorem of this chapter, Theorem 7.7, can be proved directly taking advantage of the analysis carried out there and extending it to the case p finite by means of some technicalities involving truncation operators. This will actually be our approach to pass from $p = \infty$ to finite p in Chapter 8 under the gradient constraint. Nonetheless we have chosen to proceed in a different way with the idea in mind of preparing some of the main techniques for the analysis of gradient parametrized measures pursued in Chapter 8.

The conclusion here is that there is no real restriction, except for a technical assumption, on such families of probability measures: given any family $\nu = \{\nu_x\}_{x\in\Omega}$, there always exists some sequence $\{z_j\}$ whose parametrized measure is precisely ν. The technical condition on ν is that the function of x

$$x \mapsto \int_{\mathbf{R}^m} |\lambda|^p \, d\nu_x(\lambda),$$

should be an $L^1(\Omega)$-function if we insist on having uniform $L^p(\Omega)$ bounds on our sequence. The functions z_j are assumed to take values on $\mathbf{R}^m$.

If we want to restrict further the type of sequences, various things can happen. There are examples where we obtain fundamental restrictions on the parametrized measures that such sequences can generate. This is the case of gradient parametrized measures to be discussed in the next chapter. There are instances, however, where we do not find any extra constraint on such families of probability measures. This surprising fact occurs with divergence-free parametrized measures. They will be analyzed in Chapter 10.

We need to discuss briefly some classes of functions and spaces to provide the functional analytical setting appropriate for our purposes. Let $g : \mathbf{R}^+ \to \mathbf{R}^+$ be a nondecreasing, continuous function with $\lim_{t\to\infty} g(t) = \infty$. Consider then

$$L^g(\Omega) = \left\{ z : \Omega \to \mathbf{R}^m, \text{ measurable } : \int_\Omega g(|z(x)|)\, dx < \infty \right\}.$$

Unless we demand more properties of g, $L^g(\Omega)$ may not be a vector space. We do not need this structure on $L^g(\Omega)$. For $g(t) = t^p$, $p \geq 1$ we recover the usual $L^p(\Omega)$-spaces. Similarly we set

$$\mathcal{E}^g = \left\{ \varphi : \mathbf{R}^m \to \mathbf{R}, \text{ continuous } : \lim_{|\lambda|\to\infty} \frac{\varphi(\lambda)}{1 + g(|\lambda|)} \text{ exists} \right\}.$$

It is not difficult to see that $\mathcal{E}^g$ is a separable, Banach space under the norm

$$\|\varphi\| = \left\| \frac{\varphi(\cdot)}{1 + g(|\cdot|)} \right\|_{L^\infty(\mathbf{R}^m)}.$$

As a matter of fact, $\mathcal{E}^g$ is isomorphic to $\mathcal{C}(K)$ under the sup norm where K is the one-point compactification of $\mathbf{R}^m$. The dual space $(\mathcal{E}^g)'$ strictly contains the probability measures in $\mathbf{R}^m$, μ, such that

$$\int_{\mathbf{R}^m} g(|\lambda|)\, d\mu(\lambda) < \infty.$$

There are some other objects in $(\mathcal{E}^g)'$ as well. For instance

$$\langle T, \varphi \rangle = \lim_{|\lambda|\to\infty} \frac{\varphi(\lambda)}{1 + g(|\lambda|)}$$

belongs to $(\mathcal{E}^g)'$.

If $g = +\infty$ for $t \geq R$, then the space $\mathcal{E}^g$ should be changed to

$$\mathcal{E}^g = \Big\{ \varphi : \mathbf{R}^m \to \mathbf{R}^* \cup \{+\infty\}, \text{ continuous } : \varphi = +\infty \text{ for } |\lambda| \geq R \text{ and } \lim_{|\lambda|\to R} \frac{\varphi(\lambda)}{1 + g(|\lambda|)} \text{ exists} \Big\},$$

but everything else is the same. This case should be considered in order to include the case $L^g(\Omega) = L^\infty(\Omega)$, but we do not need to make any distinction between these two cases in what follows.

2. Homogenization and localization

There are two elementary operations for analyzing parametrized measures: averaging and localization. Both processes consist in obtaining a homogeneous parametrized measure from one which is not. In the average or homogenization process, we try to somehow record in a single homogeneous parametrized measure all the information contained in all individual elements ν_x for $x \in \Omega$. While in the localization procedure, by means of a usual blow-up technique, we concentrate on a particular parametrized measure ν_a for $a \in \Omega$. We treat them succesively. The localization principle is important because it allows one to deduce properties of individual members of a family of probability measures. We will use it to derive necessary conditions in characterizing parametrized measures.

For the averaging procedure, Vitali's covering lemma enabling us to have a countable, pairwise disjoint, covering collection from any covering family of subsets is crucial to our analysis. It is also a fundamental technical tool for the proofs of characterizations of parametrized measures. A discussion of it can be found in the Appendix.

Theorem 7.1 *Let Ω and D be two regular domains in $\mathbf{R}^N$ with $|\partial\Omega| = |\partial D| = 0$. Let $\{z_j\}$ be a sequence of measurable functions over Ω, such that*

$$C = \sup_j \int_\Omega g(|z_j(x)|)\, dx < \infty,$$

for g, a continuous, nondecreasing, nonnegative function with $\lim_{t\to\infty} g(t) = \infty$. Let $\nu = \{\nu_x\}_{x\in\Omega}$ be the parametrized measure associated to some subsequence, still denoted $\{z_j\}$. There exists a sequence $\{w_j\}$ of measurable functions defined over D such that

$$\sup_j \int_D g(|w_j(x)|)\, dx < \infty,$$

and its parametrized measure is $\overline{\nu}$, homogeneous, given by

$$\langle \overline{\nu}, \varphi\rangle = \frac{1}{|\Omega|}\int_\Omega \int_{\mathbf{R}^m} \varphi(\lambda)\, d\nu_x(\lambda)\, dx.$$

Proof. The family of subsets of D given by

$$\mathcal{A}_j = \left\{a + \epsilon\overline{\Omega} \subset D : a \in D, \epsilon \le \frac{1}{j}\right\}$$

is a Vitali covering of D. There exists a countable collection $\{a_{ij} + \epsilon_{ij}\overline{\Omega}\}$, $\epsilon_{ij} \le 1/j$, pairwise disjoint and

$$D = \bigcup_i (a_{ij} + \epsilon_{ij}\overline{\Omega}) \cup N_j, \quad |N_j| = 0.$$

Notice that $\sum_i \epsilon_{ij}^N = |D| / |\Omega|$. Let us define

$$w_j(x) = z_j \left(\frac{x - a_{ij}}{\epsilon_{ij}} \right)$$

if $x \in a_{ij} + \epsilon_{ij}\overline{\Omega}$. By a natural change of variables

$$\begin{aligned}
\int_D g\left(|w_j(x)|\right) dx &= \sum_i \int_{a_{ij}+\epsilon_{ij}\Omega} g\left(\left| z_j \left(\frac{x - a_{ij}}{\epsilon_{ij}} \right) \right| \right) dx \\
&= \sum_i \epsilon_{ij}^N \int_\Omega g(|z_j(y)|)\, dy \\
&\leq C \frac{|D|}{|\Omega|} < \infty.
\end{aligned}$$

On the other hand, and using the same change of variables, if $\xi \in \mathcal{C}(\overline{D})$ and $\varphi \in \mathcal{C}_0(\mathbf{R}^m)$, we have

$$\begin{aligned}
\int_D \varphi(w_j)\xi\, dx &= \sum_i \epsilon_{ij}^N \int_\Omega \varphi(z_j(y))\xi(a_{ij} + \epsilon_{ij}y)\, dy \\
&= \sum_i \epsilon_{ij}^N\ \xi(a_{ij} + \epsilon_{ij}\overline{y}_{ij}) \int_\Omega \varphi(z_j(y))\, dy.
\end{aligned}$$

Here we have utilized the mean value theorem for integrals. We recognize in the first term a Riemann sum for the integral of ξ in D and

$$\begin{aligned}
\lim_{j\to\infty} \int_D \varphi(w_j)\xi\, dx &= \int_D \xi(x)\, dx\ \frac{1}{|\Omega|} \int_\Omega \int_{\mathbf{R}^m} \varphi(\lambda)\, d\nu(\lambda)\, dx \\
&= \int_D \xi(x)\, dx\ \langle \overline{\nu}, \varphi \rangle.
\end{aligned}$$

By the comments after the proof of Theorem 6.2, this implies that $\overline{\nu}$, homogeneous, is the parametrized measure associated to $\{w_j\}$. ■

For the localization principle that follows the Radon Nikodym and the Lebesgue Differentiation Theorem are invoked at some point.

Theorem 7.2 *Let Ω and D be as before. Let $\{z_j\}$ be such that*

$$C = \sup_j \int_\Omega g(|z_j|)\, dx < \infty, \tag{7–1}$$

where as usual, g is a continuous, nonnegative, nondecreasing function with $\lim_{t\to\infty} g(t) = \infty$. Let $\nu = \{\nu_x\}_{x\in\Omega}$ be its parametrized measure. For a.e. $a \in \Omega$ there exists a sequence $\{z_j^a\}$ defined on D such that

$$\sup_j \int_D g(|z_j^a|)\, dx < \infty,$$

and its parametrized measure is ν_a, homogeneous.

Proof. We use a blow-up argument around each point $a \in \Omega$.

Condition (7–1) enables us to affirm (through the Banach-Alaouglu-Bourbaki theorem) that

$$g(|z_j|) \overset{*}{\rightharpoonup} \mu$$

in the sense of measures so that

$$\lim_{j\to\infty} \int_\Omega \xi(x) g(|z_j|)\, dx = \int_\Omega \xi(x)\, d\mu(x),$$

for any continuous ξ, where μ is a nonnegative, finite measure. If now $\xi_{a,\rho}$ stands for a continuous function such that

$$0 \le \chi_{a+\rho D}(x) \le \xi_{a,\rho}(x) \le \chi_{a+2\rho D}(x),$$

then for some constant $M > 0$,

$$\begin{aligned}
\limsup_{\rho\to 0} \limsup_{j\to\infty} \frac{1}{\rho^N} \int_\Omega g(|z_j|)\chi_{a+\rho D}(x)\, dx \qquad & \\
\le \limsup_{\rho\to 0} \limsup_{j\to\infty} \frac{1}{\rho^N} \int_\Omega g(|z_j|)\xi_{a,\rho}(x)\, dx & \\
= \limsup_{\rho\to 0} \frac{1}{\rho^N} \int_\Omega \xi_{a,\rho}(x)\, d\mu(x) & \\
\le \limsup_{\rho\to 0} \frac{1}{\rho^N} \int_\Omega \chi_{a+2\rho D}(x)\, d\mu(x) & \\
\le M \frac{d\mu}{dx}(a). &
\end{aligned}$$

By the Radon-Nykodim theorem, the singular part of μ with respect to the Lebesgue measure is concentrated on a set of N-dimensional measure 0. Therefore $\frac{d\mu}{dx}(a) < \infty$ for a.e. $a \in \Omega$ and

$$\limsup_{\rho\to 0} \limsup_{j\to\infty} \frac{1}{\rho^N} \int_\Omega g(|z_j|)\chi_{a+\rho D}(x)\, dx < \infty, \quad \text{a.e. } a \in \Omega. \tag{7–2}$$

Define the functions

$$z^a_{j,\rho}(x) = z_j(a + \rho x), \quad x \in D, \rho > 0.$$

If $\varphi \in \mathcal{C}_0(\mathbf{R}^m)$ and $\xi \in L^\infty(D)$, we have

$$\begin{aligned}
\int_D \varphi(z^a_{j,\rho}(x))\xi(x)\, dx &= \int_D \varphi(z_j(a + \rho x))\xi(x)\, dx \\
&= \frac{1}{\rho^N} \int_\Omega \varphi(z_j(y))\chi_{a+\rho D}(y)\; \xi\left(\frac{y-a}{\rho}\right) dy.
\end{aligned}$$

Passing to the limit in j first, yields

$$\lim_{j\to\infty}\int_D \varphi(z^a_{j,\rho}(x))\xi(x)\,dx = \frac{1}{\rho^N}\int_\Omega \overline{\varphi}(y)\chi_{a+\rho D}(y)\;\xi\left(\frac{y-a}{\rho}\right)dy,$$

since $\{\varphi(z_j)\}$ converges weakly in $L^1(\Omega)$ to $\overline{\varphi}$ given by

$$\overline{\varphi}(y) = \int_{\mathbf{R}^m}\varphi(\lambda)\,d\nu_y(\lambda).$$

Next, by the Lebesgue differentiation theorem

$$\begin{aligned}\lim_{\rho\to 0}\lim_{j\to\infty}\int_D \varphi(z^a_{j,\rho}(x))\xi(x)\,dx &= \lim_{\rho\to\infty}\int_D \overline{\varphi}(a+\rho x)\xi(x)\,dx\\ &= \overline{\varphi}(a)\int_D \xi(x)\,dx,\end{aligned}$$

for a.e. $a\in D$. Due to the separability of $\mathcal{C}_0(\mathbf{R}^m)$ and $L^1(\Omega)$, we may choose a subsequence of $\{z^a_{j,\rho}\}$, which we call $\{z^a_j\}$, such that

$$\lim_{j\to\infty}\int_\Omega \xi\varphi(z^a_j)\,dx = \overline{\varphi}(a)\int_\Omega \xi\,dx,$$

for every $\varphi\in\mathcal{C}_0(\mathbf{R}^m)$ and $\xi\in L^1(\Omega)$ (by density). Since for a.e. $a\in\Omega$ and by (7–2)

$$\sup_j\int_D g(|z^a_j|)\,dx<\infty,$$

we conclude that the parametrized measure associated to the sequence $\{z^a_j\}$ is ν_a, homogeneous. ∎

In both of these theorems one could take $D=\Omega$.

3. Riemann-Lebesgue lemma

The Riemann-Lebesgue lemma is one of the most interesting nontrivial examples where we can actually determine explicitly the underlying parametrized measure associated to some sequence of functions. There are a number of different versions of it. We will concentrate on the most general of all and go into more specific versions. In some sense, this lemma is a homogenization fact and its proof is contained in the homogenization theorem, Theorem 7.1.

Lemma 7.3 *Let Ω and D be regular domains in $\mathbf{R}^N$ with $|\partial\Omega|=|\partial D|=0$ and $f\in L^p(\Omega)$. There exists a sequence $\{f_j\}$ whose associated parametrized measure is homogeneous and defined by*

$$\langle\overline{\nu},\varphi\rangle = \frac{1}{|\Omega|}\int_\Omega\varphi(f(x))\,dx,$$

for any continuous, bounded φ.

The proof of this fact is an immediate consequence of the homogenization theorem proved in the last section. For j fixed, let

$$\left\{a_{ij} + \epsilon_{ij}\Omega : \epsilon_{ij} \leq \frac{1}{j}\right\}$$

be a Vitali covering of D,

$$D = \bigcup_i (a_{ij} + \epsilon_{ij}\Omega) \cup N_j, \quad |N_j| = 0,$$

where the subsets $\{a_{ij} + \epsilon_{ij}\Omega\}$ are pairwise disjoint. Define

$$f_j(x) = f\left(\frac{x - a_{ij}}{\epsilon_{ij}}\right), \quad x \in a_{ij} + \epsilon_{ij}\Omega,$$

and conclude by the homogenization theorem.

In particular, if we take $\Omega = (0,1)^N$ and $f \in L^p(\Omega)$ extended by periodicity to all of $\mathbf{R}^N$, the sequence $z_j(x) = f(jx)$ determines the parametrized measure

$$\langle \overline{\nu}, \varphi \rangle = \frac{1}{|\Omega|} \int_\Omega \varphi(f(x))\, dx. \tag{7–3}$$

The reason why this is so is that due to the periodicity of f we can also write z_j as

$$z_j(x) = f(j(x - a_i)), \quad x \in a_i + \frac{1}{j}\Omega,$$

$$\Omega = \bigcup_{i=1}^{j^N} \left(a_i + \frac{1}{j}\Omega\right), \quad \text{certain } a_i,$$

so that $\left\{a_i + \frac{1}{j}\Omega\right\}_{i=1}^{j^N}$ is a Vitali covering of Ω. Hence the conclusion follows from Lemma 7.3.

We can specialize even more. If $f \in L^p(\Omega)$ as before with Ω the unit cube in $\mathbf{R}^N$ and extended by periodicity, for $z_j = f(jx)$ we have

$$z_j \rightharpoonup \int_\Omega f(x)\, dx \quad \text{in } L^p(\Omega),$$

taking φ as the identity in (7–3). This is what in many references is called the Riemann-Lebesgue lemma. In the very particular case in which we take $f(x) = \sin x$ or $f(x) = \cos x$ for $x \in (0, 2\pi)$, we obtain that the sequences $\{\sin(jx)\}$ and $\{\cos(jx)\}$ converge weakly in any $L^p(0, 2\pi)$ to the average of sin or cos in one period cell which is 0. This in particular implies that the Fourier coeficients of functions in $L^p(0, 2\pi)$ tend to 0.

Another interesting consequence of this lemma is that for any domain Ω and any $\lambda \in [0,1]$ we can find a sequence of characteristic functions of subsets of Ω, χ_j, such that

$$\chi_j \rightharpoonup \lambda \quad \text{in } L^p(\Omega) \text{ for any } p.$$

Take any measurable set $E \subset \Omega$ such that $|E| = \lambda\,|\Omega|$ and apply the Riemann-Lebesgue lemma to $f = \chi_E$.

4. Two auxiliary lemmas

The two basic lemmas contained in this section are important in understanding the restrictions that a given family of probability measures should verify in order to ensure that it is the parametrized measure associated with some sequence of functions in some space $L^p(\Omega)$.

For $y \in \mathbf{R}^m$ we introduce the set

$$M_y = \Big\{\nu, \text{ probability measure over } \mathbf{R}^m : \text{ there exists } z \in L^g(\Omega), \nu = \overline{\delta_{z(x)}}, \\ \int_\Omega z(x)\,dx = |\Omega|\,y\Big\}.$$

$\overline{\delta_{z(x)}}$ denotes the averaged parametrized measure associated with the constant sequence $\{z(x)\}$ according to Theorem 7.1. Recall that

$$L^g(\Omega) = \left\{z : \Omega \to \mathbf{R}^m, \text{measurable} : \int_\Omega g\left(|z(x)|\right)\,dx < \infty\right\}.$$

Note that for $\nu \in M_y$

$$\int_{\mathbf{R}^m} \varphi(\lambda)\,d\nu(\lambda) = \frac{1}{|\Omega|}\int_\Omega \varphi(z(x))\,dx.$$

Proposition 7.4 *M_y is a convex set of probability measures.*

Proof. Let $\nu_i \in M_y$, $i = 1,2$ and $\lambda \in (0,1)$. There exist $z_i \in L^g(\Omega)$, $\nu_i = \overline{\delta_{z_i(x)}}$ and $\int_\Omega z_i(x)\,dx = |\Omega|\,y$. Let us take $D \subset \Omega$, a regular subdomain with $|D| = \lambda\,|\Omega|$. By Vitali's covering lemma, there exist two countable families of subsets, $\{a_j + \epsilon_j\Omega\}$ and $\{b_j + \rho_j\Omega\}$, pairwise disjoint and

$$D = \bigcup_j (a_j + \epsilon_j\Omega) \cup N, \quad |N| = 0,$$

$$\Omega \setminus D = \bigcup_j (b_j + \rho_j\Omega) \cup N', \quad |N'| = 0.$$

We define

$$z(x) = \begin{cases} z_1\left(\dfrac{x-a_j}{\epsilon_j}\right), & x \in a_j + \epsilon_j\Omega, \\ z_2\left(\dfrac{x-b_j}{\rho_j}\right), & x \in b_j + \rho_j\Omega. \end{cases}$$

For any $\varphi \in \mathcal{E}^g$,

$$\begin{aligned} \frac{1}{|\Omega|}\int_\Omega \varphi(z(x))\,dx &= \frac{1}{|\Omega|}\sum_j \left[\epsilon_j^N \int_\Omega \varphi(z_1)\,dx + \rho_j^N \int_\Omega \varphi(z_2)\,dx\right] \\ &= \lambda\frac{1}{|\Omega|}\int_\Omega \varphi(z_1)\,dx + (1-\lambda)\frac{1}{|\Omega|}\int_\Omega \varphi(z_2)\,dx \\ &= \langle \lambda\nu_1 + (1-\lambda)\nu_2, \varphi\rangle. \end{aligned}$$

This means that $z \in L^g(\Omega)$ (taking $\varphi = g$) and $\lambda\nu_1 + (1-\lambda)\nu_2 = \overline{\delta_{z(x)}} \in M_y$. ■

Lemma 7.5 *For any continuous φ, the convexification of φ, φ^{**}, is given by*

$$\varphi^{**}(y) = \inf\left\{\int_{\mathbf{R}^m} \varphi(\lambda)\,d\nu(\lambda) : \nu \in M_y\right\}.$$

Proof. On the one hand, by Jensen's inequality

$$\varphi^{**}(y) \le \int_{\mathbf{R}^m} \varphi^{**}(\lambda)\,d\nu(\lambda) \le \int_{\mathbf{R}^m} \varphi(\lambda)\,d\nu(\lambda),$$

for $\nu \in M_y$, so that

$$\varphi^{**}(y) \le \inf\left\{\int_{\mathbf{R}^m} \varphi(\lambda)\,d\nu(\lambda) : \nu \in M_y\right\}.$$

On the other, if $y = \sum_{i=1}^n \lambda_i y_i$ with $\sum_{i=1}^n \lambda_i = 1$, $\lambda_i \ge 0$, choosing $\Omega_i \subset \Omega$ such that $|\Omega_i| = |\Omega|\,\lambda_i$, and setting

$$\sum_{i=1}^n \lambda_i\varphi(y_i) = \frac{1}{|\Omega|}\int_\Omega \varphi(z^{(n)})\,dx,$$

$$z^{(n)} = \sum_{i=1}^n \chi_{\Omega_i} y_i,$$

we have $z^{(n)} \in L^g(\Omega)$ and

$$\begin{aligned}
\varphi^{**}(y) &= \inf\left\{\sum_{i=1}^{n} \lambda_i \varphi(y_i) : \sum_{i=1}^{n} \lambda_i y_i = y, \sum_{i=1}^{n} \lambda_i = 1, \lambda_i \geq 0\right\} \\
&= \inf\left\{\frac{1}{|\Omega|}\int_\Omega \varphi(z^{(n)})\,dx : z^{(n)} = \sum_{i=1}^{n} \chi_{\Omega_i} y_i, |\Omega|\,\lambda_i = |\Omega_i|\right\} \\
&= \inf\left\{\int_{\mathbf{R}^m} \varphi(\lambda)\,d\overline{\delta_{z^{(n)}(x)}}(\lambda) : z^{(n)} = \sum_{i=1}^{n} \chi_{\Omega_i} y_i, |\Omega|\,\lambda_i = |\Omega_i|\right\} \\
&\geq \inf\left\{\int_{\mathbf{R}^m} \varphi(\lambda)\,d\nu(\lambda) : \nu \in M_y\right\}.
\end{aligned}$$

■

5. The homogeneous case

Given a family of probability measures $\nu = \{\nu_x\}_{x\in\Omega}$ depending measurably on $x \in \Omega$, when can we find a sequence of measurable functions $\{z_j\}$ such that ν is the associated parametrized measure according to Theorem 6.2? We can say that there is no real condition that ν should satisfy except for a technical assumption ensuring that the sequence of functions is bounded in $L^p(\Omega)$. We treat first the case in which ν does not depend on x, the homogeneous case, and based on this we extend the result to the nonhomogeneous case in the next section. Let again the function g be fixed.

Theorem 7.6 *Let ν be a probability measure supported in $\mathbf{R}^m$, such that*

$$\int_{\mathbf{R}^m} g(|\lambda|)\,d\nu(\lambda) < \infty.$$

There exists a sequence of functions $\{z_j\}$ such that $\{g(|z_j|)\}$ is weakly convergent in $L^1(\Omega)$ and the corresponding parametrized measure is ν, homogeneous.

Except for technical details, the proof consists in finding the sequence $\{z_j\}$ by using the Hahn-Banach theorem: the measure ν is shown to belong to the weak * closure of a convex set of measures where Dirac masses are dense. All homogeneous characterizations of parametrized measures are based on this same idea.

Proof. Let us set

$$y = \int_{\mathbf{R}^m} \lambda\,d\nu(\lambda) \in \mathbf{R}^m,$$

and consider M_y as a subset of $(\mathcal{E}^g)'$. Let T be a continuous, linear functional on $(\mathcal{E}^g)'$ under the weak * topology, so that there is a $\psi \in \mathcal{E}^g$ such that

$$\langle T, \mu\rangle = \langle \psi, \mu\rangle = \int_{\mathbf{R}^m} \psi(\lambda)\,d\mu(\lambda),$$

for $\mu \in (\mathcal{E}^g)'$. Let us suppose that T is nonnegative over M_y,

$$\int_\Omega \psi(z(x))\,dx \geq 0,$$

$$\text{for all } z \in L^g(\Omega), |\Omega|\, y = \int_\Omega z(x)\,dx.$$

Lemma 7.5 says that $\psi^{**}(y) \geq 0$ and due to Jensen's inequality

$$0 \leq \psi^{**}(y) \leq \int_{\mathbf{R}^m} \psi^{**}(\lambda)\,d\nu(\lambda) \leq \int_{\mathbf{R}^m} \psi(\lambda)\,d\nu(\lambda).$$

Thus ν cannot be separated from M_y (Hahn-Banach theorem), and by Lemma 7.4,

$$\nu \in \overline{\operatorname{co}(M_y)} = \overline{M_y}.$$

Since $\mathcal{E}^g$ is separable, there exists a sequence $\{z_j\}$ such that

$$\lim_{j\to\infty} \int_\Omega \psi(z_j)\,dx = |\Omega|\,\langle\psi,\nu\rangle, \tag{7–4}$$

for ψ in a countable, dense subset, $\mathcal{S}$, of $\mathcal{E}^g$. We can assume that $g(|\lambda|) \in \mathcal{S}$, so that $\{g(|z_j|)\}$ is uniformly bounded in $L^1(\Omega)$. By density, we can obtain that (7–4) holds for any $\psi \in \mathcal{E}^g$.

If we now apply the existence theorem, Theorem 6.2, to $\{z_j\}$ and use the averaging procedure Theorem 7.1 (observe that condition (7–4) does not change under this operation), we can assume that the parametrized measure associated to $\{z_j\}$ is homogeneous, μ. We would like to show that $\mu = \nu$. This is straightforward since for all $\varphi \in \mathcal{C}_0(\mathbf{R}^m)$ we immediately get $\langle\varphi,\nu\rangle = \langle\varphi,\mu\rangle$ as a consequence of (7–4). This identifies $\nu = \mu$.

Finally, we want to show that $\{g(|z_j|)\}$ is weakly convergent in $L^1(\Omega)$. By Chacon's biting lemma, there exists a nonincreasing sequence of measurable sets, $\{\Omega_k\}$, such that

$$g(|z_j|) \rightharpoonup \langle\, g(|\lambda|),\nu\rangle \quad \text{in } L^1(\Omega\setminus\Omega_k) \text{ for all } k.$$

Keeping in mind (7–4),

$$\begin{aligned}\lim_{k\to\infty}\lim_{j\to\infty}\int_{\Omega_k} g(|z_j|)\,dx &= \lim_{k\to\infty}\lim_{j\to\infty}\left(\int_\Omega g(|z_j|)\,dx - \int_{\Omega\setminus\Omega_k} g(|z_j|)\,dx\right)\\ &= \lim_{k\to\infty}\langle g(|\lambda|),\nu\rangle\,|\Omega_k| = 0,\end{aligned}$$

and this implies that at least for a subsequence $\{g(|z_j|)\}$ converges weakly in $L^1(\Omega)$ because the integrals of $g(|z_j|)$ are uniformly small on the exceptional sets Ω_k. ■

6. Characterization of parametrized measures

We now deal with the general, inhomogeneous case. The passage from the homogeneous case to the nonhomogeneous is done by "assembling" or patching the individual measures through the Vitali's covering technique. Although there is a considerable amount of technicality involved (especially when we place more restrictions on the sequences) the idea is simple and natural.

Theorem 7.7 *Let $\nu = \{\nu_x\}_{x\in\Omega}$ be a family of probability measures in $\mathbf{R}^m$ depending measurably on $x \in \Omega$. A necessary and sufficient condition to find a sequence of functions $\{z_j\}$ such that $\{g(|z_j|)\}$ is weakly convergent in $L^1(\Omega)$ and the associated parametrized measure is ν, is*

$$\int_\Omega \int_{\mathbf{R}^m} g(|\lambda|)\, d\nu_x(\lambda)\, dx < \infty. \tag{7–5}$$

Proof. The necessity is clear because of the representation in terms of the parametrized measure. Let us show the sufficiency. If we can find a sequence $\{z_j\}$ such that

$$\lim_{j\to\infty} \int_\Omega \xi\, \varphi(z_j)\, dx = \int_\Omega \xi(x) \int_{\mathbf{R}^m} \varphi(\lambda)\, d\nu_x(\lambda)\, dx$$

for all $\xi \in \Gamma$ and $\varphi \in \mathcal{S}$, where Γ and $\mathcal{S}$ are dense, countable subsets of $L^1(\Omega)$ and $\mathcal{C}_0(\mathbf{R}^m)$ respectively, this fact identifies $\nu = \{\nu_x\}_{x\in\Omega}$ as the parametrized measure associated to $\{z_j\}$.

Condition (7–5) implies that

$$\int_{\mathbf{R}^m} g(|\lambda|)\, d\nu_a(\lambda) < \infty,$$

for a.e. $a \in \Omega$. Let N be the complement of such a's so that $|N| = 0$. By Lemma 7.9 in the Appendix for $p = \infty$, $q = 1$ and taking $r_k(a) = 1/k$ for all $a \in \Omega \setminus N$, we have

$$\int_\Omega \xi(x)\overline{\varphi}(x)\, dx = \lim_{k\to\infty} \sum_i \overline{\varphi}(a_{ki}) \int_{a_{ki}+\epsilon_{ki}\Omega} \xi(x)\, dx \tag{7–6}$$

for all $\xi \in L^1(\Omega)$, $\varphi \in \mathcal{S}$ where

$$\Omega = \bigcup_i (a_{ki} + \epsilon_{ki}\Omega) \cup N_k, \quad |N_k| = 0,$$

$a_{ki} \in \Omega \setminus N$ and the union is pairwise disjoint. For fixed a_{ki} and by Theorem 7.6 for the homogeneous case, we can find a sequence $\left\{z_j^{ki}\right\}$ with $\nu_{a_{ki}}$ as its parametrized measure. We define then

$$z_k(x) = z_j^{ki}\left(\frac{x - a_{ki}}{\epsilon_{ki}}\right), \quad x \in a_{ki} + \epsilon_{ki}\Omega,$$

where $j = j(k,i)$ is chosen in the following way. Notice that this sequence is indexed by k rather than by j. Write $\Gamma \times \mathcal{S} = \bigcup_k D_k$, with D_k finite and $D_k \subset D_{k+1}$. For k, i fixed, choose j so that

$$\left| \epsilon_{ki}^N \int_\Omega \xi(a_{ki} + \epsilon_{ki} y)\varphi(z_j^{ki}(y))\,dy - \int_{a_{ki}+\epsilon_{ki}\Omega} \xi(x)\,dx\,\overline{\varphi}(a_{ki}) \right| \le \frac{1}{2^i k} \tag{7–7}$$

for $(\xi, \varphi) \in D_k$. This can be done precisely because $\nu_{a_{ki}}$ is generated by $\{z_j^{ki}\}$.

Let $\xi \in \Gamma$ and $\varphi \in \mathcal{S}$. If $k \ge k_0$ where $(\xi,\varphi) \in D_{k_0}$, by (7–6) and (7–7)

$$\begin{aligned} \lim_{k\to\infty} \int_\Omega \xi(x)\varphi(z_k(x))\,dx &= \lim_{k\to\infty} \sum_i \epsilon_{ki}^N \int_\Omega \xi(a_{ki} + \epsilon_{ki} y)\varphi(z_j^{ki}(y))\,dy \\ &= \lim_{k\to\infty} \sum_i \int_{a_{ki}+\epsilon_{ki}\Omega} \xi(x)\,dx\,\overline{\varphi}(a_{ki}) \\ &= \int_\Omega \xi(x)\overline{\varphi}(x)\,dx. \end{aligned} \tag{7–8}$$

For notational convenience we now change z_k to z_j. Finally, the fact that $\{g(|z_j|)\}$ is weakly convergent in $L^1(\Omega)$ can be established exactly as in the homogeneous case. By Chacon's biting lemma, there is a nonincreasing sequence of measurable sets, $\{\Omega_k\}$, such that

$$g(|z_j|) \rightharpoonup \langle g(|\lambda|), \nu_x\rangle \text{ in } L^1(\Omega \setminus \Omega_k) \text{ for all } k.$$

Because of (7–8) for $\xi \equiv 1$,

$$\begin{aligned} \lim_{k\to\infty}\lim_{j\to\infty} \int_{\Omega_k} g(|z_j|)\,dx &= \lim_{k\to\infty}\lim_{j\to\infty} \left(\int_\Omega g(|z_j|)\,dx - \int_{\Omega\setminus\Omega_k} g(|z_j|)\,dx \right) \\ &= \lim_{k\to\infty} \int_{\Omega_k} \langle g(|\lambda|), \nu_x\rangle\,dx = 0, \end{aligned}$$

because the function $\langle g(|\lambda|), \nu_x\rangle$ is an $L^1(\Omega)$-function (again due to (7–8)). ■

The particular examples we are interested in are $g(t) = t^p$ for $p > 1$ and $g(t) = +\infty$ for $t \ge R$ which corresponds to the case $p = \infty$.

We close this chapter with one interesting example. Theorem 7.7 says that any family of probability measures can be generated by an appropriate sequence of functions. Let us try to construct explicitly a generating sequence for the family of probability measures

$$\nu_x = (1-x)\delta_1 + x\delta_{-1}, \quad x \in (0,1).$$

For continuous φ, we would like to find a sequence f_j such that

$$\begin{aligned}\int_a^b \varphi(f_j(x))\,dx \to \int_a^b \int_{\mathbf{R}} \varphi(\lambda)\,d((1-x)\delta_1 + x\delta_{-1})(\lambda)\,dx \\ = \int_a^b [(1-x)\varphi(1) + x\varphi(-1)]\,dx \\ = \varphi(1)\int_a^b (1-x)\,dx + \varphi(-1)\int_a^b x\,dx.\end{aligned}$$

Let us assume that f_j takes on the values 1 and -1 in sets A_j and $B_j = (0,1)\backslash A_j$ respectively, such that $|A_j \cap [a,b]|$ is a Riemann sum for the integral of $(1-x)$ in $[a,b]$ and the same for $|B_j \cap [a,b]|$. For instance, if

$$A_j = \bigcup_{k=0}^{j} \left(\frac{k}{j}, \frac{k+1}{j} - \frac{k}{j^2}\right),$$

and we take $f_j = \chi_{A_j} - \chi_{B_j}$, then if $a_j = \frac{\langle aj\rangle}{j}$ and $b_j = \frac{\langle bj\rangle}{j}$

$$|A_j \cap [a,b]| = \frac{1}{j}\sum_{k=\langle aj\rangle}^{\langle bj\rangle} (1 - \frac{k}{j}) + |[b_j,b] \cap A_j| - |[a_j,a] \cap A_j|$$

and since $a_j \to a$ and $b_j \to b$,

$$\begin{aligned}\int_a^b \varphi(f_j)\,dx = \varphi(1)\,|A_j \cap [a,b]| + \varphi(-1)\,|B_j \cap [a,b]| \\ \to \varphi(1)\int_a^b (1-x)\,dx + \varphi(-1)\int_a^b x\,dx.\end{aligned}$$

7. Appendix

1. For a given point $x \in \mathbf{R}^m$, a sequence of sets $\{E_i\}$ shrinks suitably to x if there is $\alpha > 0$ such that each $E_i \subset B(x, r_i)$, a ball centered at x and radius $r_i > 0$, and

$$|E_i| \geq \alpha\,|B(x,r_i)|,$$

where $r_i \to 0$ as $i \to \infty$.

A family of open subsets $\{A_\lambda\}_{\lambda\in\Lambda}$ is called a Vitali covering of $\Omega \subset \mathbf{R}^m$ if for every $x \in \Omega$ there exists a sequence $\{A_i\}$ of subsets of the given family that shrink suitably to x.

Theorem 7.8 *Let $\mathcal{A} = \{A_\lambda\}_{\lambda\in\Lambda}$ be a Vitali covering of Ω. There is a sequence $\lambda_i \in \Lambda$ such that*

$$\left|\Omega \setminus \bigcup_i A_{\lambda_i}\right| = 0$$

and the subsets A_{λ_i} are pairwise disjoints.

The situation to which we apply the above covering theorem is the following. Let Ω be an open, bounded subset and B a ball containing Ω. The family of subsets

$$\mathcal{A}_k = \left\{a + \epsilon\Omega : a \in \Omega, \epsilon < \frac{1}{k}, a + \epsilon\overline{\Omega} \subset \Omega\right\}$$

is a Vitali covering of Ω. Indeed, for any $a \in \Omega$ we take $\alpha = \dfrac{|\Omega|}{|B|} > 0$ and

$$\frac{|a+\epsilon\Omega|}{|a+\epsilon B|} = \frac{|\Omega|}{|B|} = \alpha, \quad \text{for all } \epsilon.$$

Therefore by Theorem 7.8

$$\Omega = \bigcup_j (a_{jk} + \epsilon_{jk}\Omega) \bigcup N_k, \quad |N_k| = 0,$$

and the $\{a_{jk} + \epsilon_{jk}\Omega\}$ are pairwise disjoint.

2. The following is a useful, technical lemma.

Lemma 7.9 *Let $\Omega \subset \mathbf{R}^N$ be an open, bounded set with $|\partial\Omega| = 0$ and $N \subset \Omega$, a subset of measure 0. For $r_k : \Omega \setminus N \to \mathbf{R}^+$ and $\{f_j\} \subset L^p(\Omega)$, there exists a set of points $\{a_{ki}\} \subset \Omega \setminus N$ and positive numbers $\{\epsilon_{ki}\}$, $\epsilon_{ki} \le r_k(a_{ki})$ such that*

$$\left\{a_{ki} + \epsilon_{ki}\overline{\Omega}\right\} \text{ are pairwise disjoint for each } k,$$

$$\overline{\Omega} = \bigcup_i \left\{a_{ki} + \epsilon_{ki}\overline{\Omega}\right\} \cup N_k, \qquad |N_k| = 0,$$

$$\int_\Omega \xi(x) f_j(x)\,dx = \lim_{k\to\infty} \sum_i f_j(a_{ki}) \int_{a_{ki}+\epsilon_{ki}\Omega} \xi(x)\,dx$$

for every j and every $\xi \in L^q(\Omega)$, $\frac{1}{p} + \frac{1}{q} = 1$.

Proof. Let $D \subset \Omega$ be the intersection of the sets of Lebesgue points of the f_j's and set $A = \Omega \setminus N$. For each k the Lebesgue differentiation theorem implies that the family

$$F_k = \Big\{ a + \epsilon\overline{\Omega} : a \in A, \epsilon \le r_k(a), \frac{1}{|\epsilon\,\Omega|}\int_{a+\epsilon\,\Omega} |f_j(x) - f_j(a)|^p\, dx < \frac{1}{k}, \\ 1 \le j \le k, \text{and } a + \epsilon\overline{\Omega} \subset \Omega \Big\}$$

covers A in the sense of Vitali. Hence we may write

$$A = \bigcup_i \{a_{ki} + \epsilon_{ki}\overline{\Omega}\} \cup N'_k, \qquad |N'_k| = 0,$$

or

$$\overline{\Omega} = \bigcup_i \{a_{ki} + \epsilon_{ki}\overline{\Omega}\} \cup N_k, \qquad |N_k| = 0.$$

For any $\xi \in L^q(\Omega)$ and for fixed j, $k \ge j$

$$\begin{aligned}
&\left| \int_\Omega \xi(x)\ f_j(x)\, dx - \sum_i f_j(a_{ki}) \int_{a_{ki}+\epsilon_{ki}\Omega} \xi(x)\, dx \right| \\
&= \left| \sum_i \int_{a_{ki}+\epsilon_{ki}\Omega} (f_j(x) - f_j(a_{ki}))\, \xi(x)\, dx \right| \\
&\le \sum_i \left(\int_{a_{ki}+\epsilon_{ki}\Omega} |f_j(x) - f_j(a_{ki})|^p\, dx \right)^{1/p} \left(\int_{a_{ki}+\epsilon_{ki}\Omega} |\xi(x)|^q\, dx \right)^{1/q} \\
&\le \frac{1}{k} \sum_i |\epsilon_{ki}\Omega|^{1/p} \left(\int_{a_{ki}+\epsilon_{ki}\Omega} |\xi(x)|^q\, dx \right)^{1/q} \\
&\le \frac{1}{k} \left(\sum_i |\epsilon_{ki}\Omega| \right)^{1/p} \left(\sum_i \int_{a_{ki}+\epsilon_{ki}\Omega} |\xi(x)|^q\, dx \right)^{1/q} \\
&= \frac{1}{k} |\Omega|^{1/p}\, \|\xi\|_{L^q(\Omega)} \,.
\end{aligned}$$

We have used Hölder's inequality for integrals and for series. ∎

8. Bibliographical remarks

Homogenization and localization as have been presented here were introduced in [186] in the context of gradient parametrized measures to be studied in the next chapter. See also [278]. These turn out to be the basic tools that need to be redone for any new situation. The Riemann-Lebesgue lemma is a classical fact. It can also be found in [98] and textbooks dealing with Fourier Analysis.

The two auxiliary lemmas in Section 4 are well-known facts from elementary convex analysis. The rest of the chapter follows along the lines of [186] and [191]. The main facts shown in this chapter, under no differential constraint, are essentially contained as well in [93] and [302].

Concerning the technical facts in the Appendix, we refer to [283] for an accesible account on covering lemmas related to differentiation of measures. A complete treatment of Vitali's covering lemma as presented here can be found in [285]. There is a whole collection of covering lemmas that are extremely useful in Fourier Analysis. [291] is an standard reference for this topic. The last result is designed especially to be used in this book and its full proof has been included. The classical theorem of Radon-Nykodim and the Lebesgue differentiation theorem can also be found in [283].

Chapter 8
Analysis of Gradient Parametrized Measures

1. Introduction

In variational principles we are especially interested in integrands depending on gradients. For this reason we would like to study weak convergence associated to sequences of gradients. Parametrized measures associated to sequences of gradients are called gradient parametrized measures. In particular, we would like to prove a characterization for parametrized measures coming from a sequence of gradients.

The paradigm in the calculus of variations is the functional

$$I(u) = \int_\Omega \varphi(x, u(x), \nabla u(x))\, dx, \tag{8–1}$$

where Ω and φ are given, and the functions u are assumed to belong to some Sobolev class of continuous functions with weak derivatives. Let us assume we have a sequence of functions in $W^{1,p}(\Omega)$, $\{u_j\}$, converging weakly in $W^{1,p}(\Omega)$ to some $u \in W^{1,p}(\Omega)$. We know that the functions themselves converge strongly in $L^p(\Omega)$ to u, so that if we consider the parametrized measure associated to $\{(u_j, \nabla u_j)\}$ then by Proposition 6.13,

$$\mu_x = \delta_{u(x)} \otimes \nu_x, \quad \text{a.e. } x \in \Omega,$$

where $\nu = \{\nu_x\}_{x\in\Omega}$ is the parametrized measure corresponding to the sequence of gradients. We need to fix our attention on the sequence of gradients and, in particular, we may consider without loss of generality as far as understanding weak lower semicontinuity is concerned, that the integrand φ in (8–1) only depends upon the gradient variable, so that

$$I(u) = \int_\Omega \varphi(\nabla u(x))\, dx. \tag{8–2}$$

In order to obtain a characterization of parametrized measures associated to sequences of gradients there is a fundamental difference between the scalar and the vector cases which makes them drastically distinct. If we deal with functions

$$u : \Omega \subset \mathbf{R}^N \to \mathbf{R}^m$$

the scalar case corresponds to $m = 1$ whereas the vector case corresponds to $m > 1$. For such functions u, $\nabla u \in \mathbf{M}^{m\times N}$. If $m = 1$, ∇u is in fact a vector and for any two given vectors $y_1, y_2 \in \mathbf{M}^{1\times N}$ we can always find a continuous function u such that ∇u takes on only the two prescribed values y_1, y_2. This fact leads to the conclusion, after some manipulations, that if the functional I in (8–2) is weak lower semicontinuous, the integrand φ should be convex in the usual sense. In the case $m > 1$ what we have just asserted is not true: if you want to find a continuous u whose gradient ∇u takes on only two values $Y_1, Y_2 \in \mathbf{M}^{m\times N}$ then $Y_1 - Y_2$ must be a rank-one matrix. There is no way to find such a u if $Y_1 - Y_2$ does not have rank-one. This is a fact which can be proven quite simply. Take any component of u, u_i, and suppose that ∇u_i takes on two values y_1, y_2 accross a plane interface with normal vector n. Let x_0 be any point in this plane interface. $u_i(x) = y_1 \cdot x + c_1$ if $x \cdot n > 0$ for constants $y_1, c_1 \in \mathbf{R}$. Similarly $u_i(x) = y_2 \cdot x + c_2$ if $x \cdot n < 0$ for constants y_2, c_2. The continuity of u at x_0 implies $y_1x_0+c_1 = y_2x_0+c_2$ or $(y_1-y_2)\cdot x_0 = c_2-c_1$. Since this relation should hold for all x_0 in the interface we have that $(y_1 - y_2) = a_i n$, for some constant $a_i \in \mathbf{R}$. Doing this for all the components of u we see that a necessary condition to find such u is $Y_1 - Y_2 = a \otimes n$ for some vector $a \in \mathbf{R}^m$ and $n \in \mathbf{R}^N$. As we indicated in Chaper 1 this rank-one condition between Y_1 and Y_2 is also sufficient to find a continuous deformation u whose gradient takes on the two values Y_1 and Y_2.

This impossibility has the profound implication that the convexity condition can only be recovered along rank-one directions

$$\varphi(tY_1 + (1-t)Y_2) \leq t\varphi(Y_1) + (1-t)Y_2 \tag{8–3}$$

provided that $Y_1 - Y_2$ is a rank-one matrix and the functional I is weak lower semicontinuous. See Chapter 1, Section 4 for more details. Plain convexity is no longer a necessary condition for the weak lower semicontinuity of I in the vector case. New types of convexity need to be introduced. In connection with the vector case, there are three new, relevant types of convexity: polyconvexity, quasiconvexity and rank-one convexity. The latter is given in (8–3). Quasiconvexity is the necessary and sufficient condition to have weak lower semicontinuity. φ is said to be quasiconvex if

$$\varphi(Y) \leq \frac{1}{|\Omega|}\int_\Omega \varphi(Y + \nabla u(x))\, dx \tag{8–4}$$

for any matrix $Y \in \mathbf{M}^{m\times N}$ and any smooth u with compact support in Ω. If φ is quasiconvex, of all deformations having affine boundary values given by Y, φ prefers precisely the homogeneous, affine deformation determined by Y.

Polyconvexity provides sufficient conditions for quasiconvexity since (8–4), which is hard to understand in practice.

It turns out that parametrized measures generated by gradients are characterized by Jensen's inequality for all quasiconvex functions (more precise statements are needed concerning growth restrictions). The technique to show this result is essentially the same used in the previous chapter for the general case. As a matter of fact, we will try along this chapter to reproduce the same tools for the gradient case, point out where they fail and how difficulties can be overcome. Quasiconvexity will be closely examined along the way and some of its basic properties established.

Throughout this chapter the function g will be taken to be $g(t) = t^p$ for $1 \leq p < \infty$ or $g(t) = +\infty$ for $t \geq R$, some positive R, so that we will concentrate on $L^p(\Omega)$-bounds for sequences of gradients. In these cases we put $\mathcal{E}^p$ and $(\mathcal{E}^p)'$ for the appropriate spaces of functions to be considered and their duals:

$$\mathcal{E}^p = \left\{ \varphi : \mathbf{M}^{m\times N} \to \mathbf{R}, \text{ continuous } : \lim_{|A|\to\infty} \frac{\varphi(A)}{1+|A|^p} \text{ exists} \right\}.$$

We will also keep in mind the nonseparable spaces

$$X^p = \left\{ \varphi : \mathbf{M}^{m\times N} \to \mathbf{R}, \text{ continuous} : |\varphi(A)| \leq C(1+|A|^p), C \in \mathbf{R} \right\}.$$

$\mathcal{E}^p$ is a close subspace of X^p under the same norm

$$\|\varphi\| = \left\| \frac{\varphi(\cdot)}{1+|\cdot|^p} \right\|_{L^\infty(\mathbf{M}^{m\times N})}.$$

The dual space $(X^p)'$ can indeed be identified with the set of finite, Radon measures over $\mathbf{M}^{m\times N}$, μ, such that $|\cdot|^p$ belongs to $L^1(d\mu)$.

For the sake of brevity in some of our statements, we agree in calling a family of probability measures $\nu = \{\nu_x\}_{x\in\Omega}$ supported in $\mathbf{M}^{m\times N}$, a $W^{1,p}$-parametrized measure if it can be generated by the gradients of a bounded sequence of functions in $W^{1,p}(\Omega)$. We include here the case $p = \infty$. This is the same terminology introduced in Chapter 1.

2. Homogenization, localization and the Riemann-Lebesgue lemma

To begin with we examine the homogenization and localization procedures and the Riemann-Lebesgue lemma when working with gradients: the sequence z_j is now ∇u_j for $u_j \in W^{1,p}(\Omega)$. In this section we do not need to distinguish between the cases $1 \leq p < \infty$ and $p = \infty$, or between the scalar and the vector cases.

Let us look first at the homogenization process. Going back to the proof of Theorem 7.1, we realize that if we can find a sequence of functions $w_j \in W^{1,p}(\Omega)$ such that

$$\nabla w_j(x) = \nabla u_j\left(\frac{x - a_{ij}}{\epsilon_{ij}}\right), \tag{8–5}$$

for $x \in a_{ij} + \epsilon_{ij}\Omega$, where once again by the Vitali's covering lemma

$$\Omega = \bigcup_i (a_{ij} + \epsilon_{ij}\overline{\Omega}) \cup N_j, \quad |N_j| = 0,$$

then the proof is exactly the same as that of Theorem 7.1. Condition (8–5) can be fulfilled by simply putting

$$w_j(x) = \epsilon_{ij}\, u_j\left(\frac{x - a_{ij}}{\epsilon_{ij}}\right),$$

for $x \in a_{ij} + \epsilon_{ij}\Omega$. It may not be true though that w_j so defined belongs to $W^{1,p}(\Omega)$. Indeed, we know that w_j should be much more regular than simply belonging to $L^p(\Omega)$. For p sufficiently large, w_j ought to be even continuous, and this might not be the case if we are not more careful about our definition of w_j. The only known way of having the continuity property for w_j is to enforce affine boundary values for u_j. Let $Y \in \mathbf{M}^{m\times N}$ and let u_Y be the affine function $u_Y(x) = Yx$. If $u_j \in W^{1,p}(\Omega)$ and $u_j - u_Y \in W_0^{1,p}(\Omega)$ the function

$$w_j(x) = \begin{cases} \epsilon_{ij}\, u_j\left(\dfrac{x - a_{ij}}{\epsilon_{ij}}\right) + u_Y(a_{ij}), & x \in a_{ij} + \epsilon_{ij}\Omega, \\ u_Y(x), & \text{otherwise}, \end{cases}$$

is well defined as a function in $W^{1,p}(\Omega)$ since it is continuous (easy to check), $w_j - u_Y \in W_0^{1,p}(\Omega)$ and its gradient satisfies (8–5). Hence we have proved the following theorem.

Theorem 8.1 *Let $\{u_j\}$ be a bounded sequence of functions in $W^{1,p}(\Omega)$ with affine boundary values given by u_Y. Let $\nu = \{\nu_x\}_{x\in\Omega}$ be the parametrized measure associated with $\{\nabla u_j\}$. There exists a sequence $\{w_j\}$, bounded in $W^{1,p}(\Omega)$ with the same boundary values, such that the corresponding parametrized measure is $\overline{\nu}$, homogeneous, and given by*

$$\langle \overline{\nu}, \varphi \rangle = \frac{1}{|\Omega|}\int_\Omega \int_{\mathbf{R}^m} \varphi(\lambda)\, d\nu_x(\lambda)\, dx.$$

Since the Riemann-Lebesgue lemma is a homogenization fact, its proof for the gradient case is based in the same idea. We need to have the precaution enforcing affine boundary values.

Lemma 8.2 *Let Ω and D be two domains in $\mathbf{R}^N$ (open, bounded, regular subsets) and $u \in W^{1,p}(D)$, $u - u_Y \in W_0^{1,p}(D)$. There exists a sequence $\{w_j\}$ bounded in $W^{1,p}(\Omega)$, $w_j - u_Y \in W_0^{1,p}(\Omega)$, such that the parametrized measure associated with the sequence $\{\nabla w_j\}$ is homogeneous and given by*

$$\langle \overline{\nu}, \varphi \rangle = \frac{1}{|D|} \int_D \varphi(\nabla u(x))\, dx,$$

for any $\varphi \in X^p$.

The proof is again a basic application of the Vitali's covering lemma. For j fixed, let

$$\left\{ a_{ij} + \epsilon_{ij} D : \epsilon_{ij} \le \frac{1}{j} \right\}$$

be a Vitali covering of Ω,

$$\Omega = \bigcup_i (a_{ij} + \epsilon_{ij} D) \cup N_j, \quad |N_j| = 0,$$

where the subsets $\{a_{ij} + \epsilon_{ij} D\}$ are pairwise disjoint. Define

$$w_j(x) = \begin{cases} \epsilon_{ij}\, u\left(\dfrac{x - a_{ij}}{\epsilon_{ij}}\right) + u_Y(a_{ij}), & x \in a_{ij} + \epsilon_{ij}\Omega, \\ u_Y(x), & \text{otherwise.} \end{cases}$$

This lemma will prove to be important in understanding later on the concept of quasiconvexity from the point of view of parametrized measures.

The other versions of this lemma previously discussed have also a counterpart with periodic gradients. They are left to the reader.

We now turn to the localization property. Once more the proof is analogous to the general case provided we define properly the new sequence of functions giving rise to the new parametrized measure. In this case we should have

$$\nabla u^a_{j,\rho}(x) = \nabla u_j(a + \rho x), \quad \rho > 0.$$

As in Theorem 7.2, the sequence $\{\nabla u^a_{j,\rho}\}$ is uniformly bounded in $L^p(\Omega)$ in j and ρ for a.e. $a \in \Omega$, and if $\nu = \{\nu_x\}_{x\in\Omega}$ is the parametrized measure associated to $\{\nabla u_j\}$, then $\{\nabla u^a_{j,\rho}\}$ generates ν_a, homogeneous, for an appropriate subsequence. In order to have a bounded sequence of $W^{1,p}(\Omega)$-functions, $\{u^a_{j,\rho}\}$, as $\rho \to 0$, define

$$F(x) = \int_{\mathbf{M}^{m\times N}} A\, d\nu_x(A),$$

the linear function $u_a(x) = F(a)x$, for $a \in \Omega$, and its average over Ω, $C_a = \frac{1}{|\Omega|} \int_\Omega u_a(x)\, dx$. Take

$$u^a_{j,\rho}(x) = \frac{1}{\rho}(u_j(a + \rho x) - M_{a,j,\rho}), \quad x \in \Omega, \tag{8–6}$$

where the constant $M_{a,j,\rho}$ is chosen so that

$$\int_\Omega u^a_{j,\rho}(x)\,dx = C_a.$$

By Poincaré's inequality, the sequence $\{u^a_{j,\rho}\}$ is truly bounded in $W^{1,p}(\Omega)$ independently of j and ρ and for almost every $a \in \Omega$.

We would also like to incorporate the affine boundary values for the new sequence defining the localized, homogeneous parametrized measure at almost every point in Ω. In order to do that we have the following lemma.

Lemma 8.3 *Let $\{v_j\}$ be a bounded sequence in $W^{1,p}(\Omega)$ such that the sequence $\{\nabla v_j\}$ generates the parametrized measure $\nu = \{\nu_x\}_{x\in\Omega}$. Let*

$$\nabla u(x) = \int_{\mathbf{M}^{m\times N}} A\,d\nu_x(A) \in \mathbf{M}^{m\times N}, \quad u \in W^{1,p}(\Omega),$$

so that $v_j \rightharpoonup u$ in $W^{1,p}(\Omega)$. There exists a new sequence $\{u_k\}$, bounded in $W^{1,p}(\Omega)$, such that $\{\nabla u_k\}$ generates the same parametrized measure ν and $u_k - u \in W^{1,p}_0(\Omega)$ for all k. If for $p < \infty$ $\{|\nabla v_j|^p\}$ is equiintegrable, so is $\{|\nabla u_k|^p\}$.

Proof. Let $\{\eta_k\}$ be a sequence of cut-off functions with the properties:

i) $\eta_k = 1$ over $\partial\Omega$;

ii) $\eta_k \equiv 0$ in $\Omega_k = \{x \in \Omega : \mathrm{dist}(x, \partial\Omega) \geq 1/k\}$;

iii) $|\nabla \eta_k| \leq C/k$, for some constant C.

We consider the functions

$$w_{jk}(x) = \eta_k(x)u(x) + (1 - \eta_k(x))v_j(x).$$

Clearly $w_{jk} - u \in W^{1,p}_0(\Omega)$ for all j, k. Moreover

$$\nabla w_{jk} = \eta_k(x)\nabla u(x) + (1 - \eta_k(x))\nabla v_j(x) + (u(x) - v_j(x)) \otimes \nabla\eta_k(x).$$

Since $\|v_j - u\|_{L^p(\Omega)}$ converges to 0, a subsequence $j(k)$ can be chosen so that the third term in the above sum tends to 0 and it is therefore weakly convergent in $L^1(\Omega)$. The first term is also weakly convergent. The middle term in the sum is also bounded or weakly convergent in $L^1(\Omega)$. Let $\{u_k\}$ be the sequence $\{w_{j(k)k}\}$, so that $\{|\nabla u_k|^p\}$ is either bounded or weakly convergent in $L^1(\Omega)$, $u_k - u \in W^{1,p}_0(\Omega)$. The fact

$$\left|\{\nabla v_j \neq \nabla w_{j(k)k}\}\right| = |\{\eta_k < 1\}| \to 0, \quad k \to \infty,$$

implies by Lemma 6.3 that the parametrized measure associated to $\{\nabla u_k\}$ is again ν. ∎

If we apply the lemma to the subsequence of $\{u^a_{j,\rho}\}$ given in (8–6) we obtain our version of the localization property for gradients.

Theorem 8.4 *Let $\{u_j\}$ be a bounded sequence in $W^{1,p}(\Omega)$ and $\nu = \{\nu_x\}_{x\in\Omega}$ the parametrized measure associated to $\{\nabla u_j\}$. Let*

$$F(a) = \int_{\mathbf{M}^{m\times N}} A\, d\nu_a(A)$$

and $u_a(x) = F(a)x$, $a \in \Omega$. For a.e. $a \in \Omega$, there exists a sequence $\{w^a_j\}$ bounded in $W^{1,p}(\Omega)$ such that $w^a_j - u_a \in W^{1,p}_0(\Omega)$, for all j and the parametrized measure associated to $\{\nabla w^a_j\}$ is ν_a, homogeneous.

3. The scalar case

Let us not make any distinction between the scalar and the vector case for the moment and try to reproduce the lemmas in Section 4 of the previous chapter in the special case when we are working with gradients. In this situation gradients are matrices in $\mathbf{M}^{m\times N}$. We define the analogue of M_y by

$$\begin{aligned} M_Y = \Big\{\nu, \text{ probability measures over } \mathbf{M}^{m\times N} : \nu = \overline{\delta_{\nabla u(x)}}, u \in W^{1,p}(\Omega), \\ u - u_Y \in W^{1,p}_0(\Omega)\Big\} \subset (\mathcal{E}^p)', \end{aligned}$$

where $Y \in \mathbf{M}^{m\times N}$, u_Y is the linear transformation Yx and $\overline{\nu_x}$ represents the averaged parametrized measure for ν_x as discussed in Section 2. The reason why we impose a boundary condition on u rather than an average condition should be clear after the discussion of the averaging procedure.

Lemma 8.5 *M_Y is convex.*

Proof. Let $u_i \in W^{1,p}(\Omega)$, $u_i - u_Y \in W^{1,p}_0(\Omega)$, $i = 1, 2$ and let ν_i be the average of the gradients of u_i as in the definition of M_Y. Let $\lambda \in (0,1)$. Take a regular, open subdomain $D \subset \Omega$ such that $|D| = \lambda|\Omega|$. There exist two countable families of subsets of D and $\Omega \setminus D$ of the type

$$\{a_i + \epsilon_i\Omega : a_i \in D, \epsilon_i > 0, a_i + \epsilon_i\Omega \subset D\}$$
$$\{b_i + \rho_i\Omega : b_i \in \Omega\setminus D, \rho_i > 0, b_i + \rho_i\Omega \subset \Omega\setminus D\}$$

such that

$$D = \bigcup_i (a_i + \epsilon_i\Omega) \cup N, \quad |N| = 0,$$

$$\Omega\setminus D = \bigcup_i (b_i + \rho_i\Omega) \cup N', \quad |N'| = 0.$$

Define u by

$$u(x) = \begin{cases} \epsilon_i \, u_1\left(\frac{x-a_i}{\epsilon_i}\right) + u_Y(a_i), & x \in a_i + \epsilon_i \Omega, \\ \rho_i \, u_2\left(\frac{x-b_i}{\rho_i}\right) + u_Y(b_i), & x \in b_i + \rho_i \Omega, \\ u_Y(x), & \text{otherwise.} \end{cases}$$

$u - u_Y \in W_0^{1,p}(\Omega)$ and

$$\nabla u(x) = \begin{cases} \nabla u_1\left(\frac{x-a_i}{\epsilon_i}\right), & x \in a_i + \epsilon_i \Omega, \\ \nabla u_2\left(\frac{x-b_i}{\rho_i}\right), & x \in b_i + \rho_i \Omega. \end{cases}$$

As in the proof of the analogous lemma in the previous chapter, for $\varphi \in \mathcal{E}^p$ we get

$$\begin{aligned} \int_\Omega \varphi(\nabla u(x))\, dx &= \sum_i \left[\epsilon_i^N \int_\Omega \varphi(\nabla u_1)\, dx + \rho_i^N \int_\Omega \varphi(\nabla u_2)\, dx \right] \\ &= \lambda \int_\Omega \varphi(\nabla u_1)\, dx + (1-\lambda) \int_\Omega \varphi(\nabla u_2)\, dx \\ &= \langle \lambda \nu_1 + (1-\lambda)\nu_2, \varphi \rangle. \end{aligned}$$

In particular $u \in W^{1,p}(\Omega)$ and $\lambda\nu_1 + (1-\lambda)\nu_2 \in M_Y$. ■

In trying to prove the next lemma we see the basic difference between the scalar case and the vector case. Let us take first $m = 1$.

Lemma 8.6 *Let $m = 1$ and let φ be any continuous function. Then*

$$\varphi^{**}(Y) = \inf \left\{ \int_{\mathbf{M}^{1\times N}} \varphi(A)\, d\nu(A) : \nu \in M_Y \right\}.$$

Proof. Let us assume that we have $Y = tY_1 + (1-t)Y_2$ where $Y, Y_1, Y_2 \in \mathbf{M}^{1\times N}$. They are in fact vectors. Consider the function

$$u(x) = Y_2 x + \int_0^{(Y_1 - Y_2)\cdot x} \chi_t(s)\, ds.$$

χ_t is the characteristic function of the interval $(0,t)$ in $(0,1)$ extended by periodicity to all of $\mathbf{R}$. We notice that

$$\begin{aligned} \nabla u(x) &= Y_2 + \chi_t((Y_1 - Y_2)\cdot x)(Y_1 - Y_2) \\ &= \begin{cases} Y_1, & (Y_1 - Y_2)\cdot x - \langle (Y_1 - Y_2)\cdot x \rangle < t, \\ Y_2, & (Y_1 - Y_2)\cdot x - \langle (Y_1 - Y_2)\cdot x \rangle > t. \end{cases} \end{aligned}$$

Define $u_j(x) = (1/j)u(jx)$. We know that the associated parametrized measure is

$$\nu = t\delta_{Y_1} + (1-t)\delta_{Y_2}$$

homogeneous. By Lemma 8.3 $\nu \in \overline{M_Y}$ (closure meant in the weak * topology) and

$$t\varphi(Y_1) + (1-t)\varphi(Y_2) \geq \inf\left\{\int_{\mathbf{M}^{1\times N}} \varphi(A)\, d\nu(A) : \nu \in M_Y\right\}.$$

Consider now the situation in which $Y = \sum_{i=1}^4 \lambda_i Y_i$, a convex combination. Renaming and rearranging this sum we could write

$$Y = tY^{(1)} + (1-t)Y^{(2)},$$
$$Y^{(1)} = t^{(1)}Y_1^{(1)} + (1-t^{(1)})Y_2^{(1)},$$
$$Y^{(2)} = t^{(2)}Y_1^{(2)} + (1-t^{(2)})Y_2^{(2)},$$

for appropriate $t, t^{(1)}, t^{(2)}$. Applying the preceding arguments to $Y, Y^{(1)}, Y^{(2)}$ we can find a sequence $\{v_k\}$ whose gradients take on essentially the values $Y^{(1)}$ and $Y^{(2)}$ with relative frequency t and $1-t$ respectively. For k and $i = 1, 2$ fixed, let Ω_k^i stands for the set where $\nabla v_k = Y^{(i)}$ and replace Ω by Ω_k^i in the previous construction for a convex combination of just two matrices corresponding to the decomposition $Y^{(i)} = t^{(i)}Y_1^{(i)} + (1-t^{(i)})Y_2^{(i)}$, $i = 1, 2$. We obtain sequences $\{v_j^{ik}\}$ whose gradients take on the values, except on small sets $E_j^{ik} \subset \Omega_k^i$, $Y_1^{(i)}$ and $Y_2^{(i)}$ with relative frequency $t^{(i)}$ and $1-t^{(i)}$, respectively. Choose $j = j(i,k)$ large enough so that

$$\frac{|E_j^{ik}|}{|\Omega_k^i|} \to 0$$

as $k \to \infty$ uniformly in $i = 1, 2$ and define

$$u_k(x) = \begin{cases} v_j^{ik}(x), & \text{if } x \in \Omega_k^i, \\ v_k(x), & \text{else.} \end{cases}$$

These functions u_k are well defined as $W^{1,p}(\Omega)$-functions since we are keeping track of the boundary values on $\partial\Omega_k^i$. It is an interesting exercise to realize that if ν_k is the average of the gradient of u_k, by the constructions carried out,

$$\int_{\mathbf{M}^{1\times N}} \varphi(A)\, d\nu_k(A) \to$$
$$t(t^{(1)}\varphi(Y_1^{(1)}) + (1-t^{(1)})\varphi(Y_2^{(1)})) + (1-t)(t^{(2)}\varphi(Y_1^{(2)}) + (1-t^{(2)})\varphi(Y_2^{(2)})),$$

as $k \to \infty$. Since each $\nu_k \in M_Y$, because the boundary condition is kept throughout the above process,

$$t(t^{(1)}\varphi(Y_1^{(1)}) + (1-t^{(1)})\varphi(Y_2^{(1)})) + (1-t)(t^{(2)}\varphi(Y_1^{(2)}) + (1-t^{(2)})\varphi(Y_2^{(2)})) \geq$$
$$\inf\left\{\int_{\mathbf{M}^{1\times N}} \varphi(A)\, d\nu(A) : \nu \in M_Y\right\}.$$

Generalizing this construction recursively for any finite number of vectors Y_i and numbers t_i and using Jensen's inequality, we get

$$\begin{aligned}\varphi^{**}(Y) &= \inf\left\{\sum_{i=1}^{n} t_i\varphi(Y_i) : t_i > 0, \sum_{i+1}^{n} t_i = 1, \sum_{i=1}^{n} t_i Y_i = Y\right\}\\ &\geq \inf\left\{\int_{\mathbf{M}^{1\times N}} \varphi(A)\, d\nu(A) : \nu \in M_Y\right\}\\ &\geq \inf\left\{\int_{\mathbf{M}^{1\times N}} \varphi^{**}(A)\, d\nu(A) : \nu \in M_Y\right\}\\ &\geq \varphi^{**}(Y).\end{aligned}$$

Notice that for any $u \in W^{1,p}(\Omega)$ such that $u - u_Y \in W_0^{1,p}(\Omega)$ it is true that if $\nu = \overline{\delta_{\nabla u}}$, by the divergence theorem

$$\int_{\mathbf{M}^{1\times N}} A\, d\nu(A) = \frac{1}{|\Omega|}\int_\Omega \nabla u(x)\, dx = Y.$$

The proof is finished. ∎

With these two lemmas at hand one can reproduce almost exactly the proofs of Theorems 7.6 and 7.7, and obtain the corresponding characterization for the scalar case when $m = 1$. There is an important difference in the fact that we are working with gradients so that proofs change somewhat. Since we will redo these proofs for the vector case in subsequent sections, we refer the reader to those sections for the proofs.

Theorem 8.7 *Let $\nu = \{\nu_x\}_{x\in\Omega}$ be a family of probability measures supported in $\mathbf{M}^{1\times N}$. The necessary and sufficient conditions to find a bounded sequence of functions, $\{u_j\}$, in $W^{1,p}(\Omega)$, such that for $1 < p < \infty$, $\{|\nabla u_j|^p\}$ is weakly convergent in $L^1(\Omega)$, and the parametrized measure associated to $\{\nabla u_j\}$ is ν, are*

$$\nabla u(x) = \int_{\mathbf{M}^{1\times N}} A\, d\nu_x(A), \qquad \text{for some } u \in W^{1,p}(\Omega),$$

and

$$\begin{aligned}\int_\Omega\int_{\mathbf{M}^{1\times N}} |A|^p\, d\nu_x(A)\, dx < \infty, &\quad 1 \leq p < \infty,\\ \operatorname{supp}(\nu_x) \subset K, \text{ a.e. } x \in \Omega, &\quad p = \infty,\end{aligned}$$

where $K \subset \mathbf{M}^{1\times N}$ is a fixed compact set.

This theorem is not valid for $p = 1$. In this case, even though $\{v_j\}$ is a bounded sequence in $W^{1,1}(\Omega)$ it may not be true that $v_j \rightharpoonup v$ in $W^{1,1}(\Omega)$ with $v \in W^{1,1}(\Omega)$ so that there is no way to know if the first moment $\int_{\mathbf{M}} A\, d\nu_x(A)$ is the gradient of a $W^{1,1}(\Omega)$-function. If we explicitly assume this as fact, the theorem is valid for $p = 1$ as well.

4. Quasiconvexity

Let us now deal with the vector case $m > 1$ and take $p = \infty$ for the rest of this section. In this case the set M_Y is a set of probability measures compactly supported in $\mathbf{M}^{m\times N}$ since the gradients of functions in $W^{1,\infty}(\Omega)$ are bounded. We simply set

$$M_Y = \Big\{\nu, \text{ probability measure } : \nu = \overline{\delta_{\nabla u(x)}}, u \in W^{1,\infty}(\Omega), \\ u - u_Y \in W_0^{1,\infty}(\Omega)\Big\}.$$

Assume we would like to redo the proof of Lemma 8.6 in this case. Given two matrices we need to find a continuous function u such that its gradient ∇u takes on two values $Y_1, Y_2 \in \mathbf{M}^{m\times N}$. The continuity of u is necessary because if such u is to exist its gradient would be bounded and therefore $u \in W^{1,\infty}(\Omega)$. This condition places an important restriction on Y_1, Y_2. In the Introduction and in Chapter 1 we have argued that a necessary and sufficient condition for finding such a Lipschitz function u is $Y_1 - Y_2 = a \otimes n$ for some vector $a \in \mathbf{R}^m$ and $n \in \mathbf{R}^N$. Indeed, if $Y_1 - Y_2 = a \otimes n$ then we can take

$$u(x) = Y_2 x + \int_0^{n\cdot x} \chi_t(s)\, ds\ a.$$

In this case

$$\begin{aligned}\nabla u(x) =& Y_2 + \chi_t(n\cdot x) a \otimes n \\ =& \begin{cases} Y_1, & n\cdot x - \langle n \cdot x\rangle < t, \\ Y_2, & n\cdot x - \langle n \cdot x\rangle > t,\end{cases}\end{aligned}$$

and we could redo Lemma 8.6. The condition $Y_1 - Y_2 = a \otimes n$ places a real and fundamental restriction so that in general the infimum

$$\inf\left\{\int_{\mathbf{M}^{m\times N}} \varphi(A)\, d\nu(A) : \nu \in M_Y\right\} \tag{8–7}$$

is no longer $\varphi^{**}(Y)$. For example, take $m = N = 2$ and let $\varphi(A) = \det A$.

Lemma 8.8 *Let $u \in W^{1,\infty}(\Omega)$ such that $u - u_Y \in W_0^{1,\infty}(\Omega)$. Then*

$$\int_\Omega \det(\nabla u(x))\, dx = |\Omega| \det Y.$$

Proof. Let φ be a $\mathcal{C}_0^\infty(\Omega)$ and consider the function $v(x) = u_Y(x) + \varphi(x)$. We claim that

$$\int_\Omega \det(\nabla v)\, dx = |\Omega| \det Y.$$

In fact

$$\det(\nabla v) = \det(Y + \nabla\varphi) = \det Y + \det \nabla\varphi - (\text{adj}Y)^T \cdot \nabla\varphi.$$

The third term on the right-hand side involves some partial derivative of some component of φ multiplied by some component of Y. Since $\varphi \in \mathcal{C}_0^\infty(\Omega)$, by the divergence theorem the integral over Ω of these terms vanish. Concerning the second term it is very easy to obtain

$$\det(\nabla\varphi) = \text{div}\left(-\varphi_2\frac{\partial\varphi_1}{\partial x_2}, \varphi_2\frac{\partial\varphi_1}{\partial x_1}\right).$$

Again by the divergence theorem we have that the integral of this term also vanishes. The result follows approximating $u - u_Y$ by a sequence of smooth functions in the strong $W^{1,\infty}(\Omega)$-topology (observe that det is continuous in the strong topology of $W^{1,\infty}(\Omega)$). ■

For $\nu \in M_Y$ there exists $u \in W^{1,\infty}(\Omega)$, $u - u_Y \in W_0^{1,\infty}(\Omega)$ and by the lemma

$$\int_{\mathbf{M}^{2\times 2}} \det A\, d\nu(A) = \frac{1}{|\Omega|}\int_\Omega \det(\nabla u(x))\, dx = \det Y.$$

This means that in fact

$$\det(Y) = \inf\left\{\int_{\mathbf{M}^{2\times 2}} \det(A)\, d\nu(A) : \nu \in M_Y\right\}$$

even though det is not convex as can be checked very easily.

Given a continuous function φ defined over matrices, the above infimum (8–7) is called the quasiconvexification of φ at Y and is denoted by $Q\varphi(Y)$. Equivalently

$$Q\varphi(Y) = \frac{1}{|\Omega|}\inf\left\{\int_\Omega \varphi(\nabla u(x))\, dx : u \in W^{1,\infty}(\Omega), u - u_Y \in W_0^{1,\infty}(\Omega)\right\}.$$

As we will show shortly this definition does not depend upon Ω: the infimum is the same for all regular domains. We say that a continuous function φ is quasiconvex if it coincides with its quasiconvexification for every matrix Y, i.e.

$$\varphi(Y) = \frac{1}{|\Omega|}\inf\left\{\int_\Omega \varphi(\nabla u(x))\, dx : u \in W^{1,\infty}(\Omega), u - u_Y \in W_0^{1,\infty}(\Omega)\right\},$$

for all Y. In particular, det is a quasiconvex function on 2×2 matrices. An equivalent way of declaring a function φ quasiconvex is by requiring

$$\varphi(Y) \leq \frac{1}{|\Omega|} \int_{\Omega} \varphi(Y + \nabla u(x))\, dx, \tag{8–8}$$

for all $u \in W_0^{1,\infty}(\Omega)$.

As pointed out, we could repeat the proof of Lemma 8.6 if we have the condition $Y_1 - Y_2 = a \otimes n$ and this condition is also necessary if at any time we would like to obtain a $W^{1,\infty}$-function in the appropriate domain whose gradient takes on the values Y_1 and Y_2 across a plane interface with normal n. To emphasize this idea, a function φ is called rank-one convex if

$$\varphi(tY_1 + (1-t)Y_2) \leq t\varphi(Y_1) + (1-t)\varphi(Y_2), \quad 0 \leq t \leq 1,$$

whenever $Y_1 - Y_2 = a \otimes n$ is a rank-one matrix. It should be clear that if φ has this rank-one convexity property, the value $\varphi(Y)$ would not be lowered by the processes described in the proof of Lemma 8.6. As a consequence, if φ is quasiconvex then it also is rank-one convex. The converse is not true. Said differently, it is not the case in general that the infimum in (8–7) is obtained by the scheme described in the proof of Lemma 8.6. We will examine in detail this question in Chapter 9.

We now prove four basic properties of quasiconvexity. The first lemma establishes that the quasiconvexification is well defined and does not depend on the particular regular domain Ω that we use in its definition.

Lemma 8.9 *Let Ω and Ω' be two open, bounded subsets of $\mathbf{R}^N$ with $|\partial\Omega| = |\partial\Omega'| = 0$, and let φ be a function defined on $\mathbf{M}^{m\times N}$. For any matrix Y, the two infima*

$$\frac{1}{|\Omega|} \inf \left\{ \int_{\Omega} \varphi(\nabla u(x))\, dx : u \in W^{1,\infty}(\Omega), u - u_Y \in W_0^{1,\infty}(\Omega) \right\},$$

and

$$\frac{1}{|\Omega'|} \inf \left\{ \int_{\Omega'} \varphi(\nabla v(x))\, dx : v \in W^{1,\infty}(\Omega'), v - u_Y \in W_0^{1,\infty}(\Omega') \right\}$$

are equal.

Proof. By Vitali's covering lemma, we can write

$$\Omega' = \bigcup_i (a_i + \epsilon_i \Omega) \cap N, \quad |N| = 0,$$

where $a_i \in \Omega'$, $\epsilon_i > 0$, $a_i + \epsilon_i\Omega \subset \Omega'$ for all i. Observe that $|\Omega'| = \sum_i \epsilon_i^N |\Omega|$. For any given $u \in W^{1,\infty}(\Omega)$ such that $u - u_Y \in W_0^{1,\infty}(\Omega)$, consider

$$v(y) = \begin{cases} \epsilon_i\, u\left(\dfrac{y-a_i}{\epsilon_i}\right) + u_Y(a_i), & \text{if } y \in a_i + \epsilon_i\Omega, \\ u_Y(y), & \text{otherwise.} \end{cases}$$

It is easily verified that $v \in W^{1,\infty}(\Omega')$, $v - u_Y \in W_0^{1,\infty}(\Omega')$ and

$$\begin{aligned} \int_{\Omega'} \varphi(\nabla v(y))\, dy &= \sum_i \int_{a_i+\epsilon_i\Omega} \varphi(\nabla u(\frac{y-a_i}{\epsilon_i})\, dy \\ &= \sum_i \epsilon_i^N \int_\Omega \varphi(\nabla u(x))\, dx \\ &= \frac{|\Omega'|}{|\Omega|} \int_\Omega \varphi(\nabla u(x))\, dx. \end{aligned}$$

If we interchange the roles of Ω and Ω' we obtain the desired result. ∎

The second property states that the quasiconvexification of a continuous function φ is indeed quasiconvex as one would reasonably expect.

Lemma 8.10 *Let φ be an upper-semicontinuous function defined on $\mathbf{M}^{m\times N}$. The quasiconvexification of φ is a quasiconvex function.*

Proof. We proceed in several steps.

Step 1. Let u be a piecewise affine function such that $u - u_Y \in W_0^{1,\infty}(\Omega)$ for some fixed matrix Y. We can set

$$\nabla u = \sum_i Y_i \chi_{\Omega_i},$$

$$u(x) = a_i + F_i x, \quad x \in \Omega_i, a_i \in \mathbf{R}^m,$$

$$\Omega = \bigcup_i \Omega_i, \quad \Omega_i, \text{ pairwise disjoint open subdomains.}$$

Then

$$\begin{aligned} \int_\Omega Q\varphi(\nabla u)\, dx &= \sum_i |\Omega_i|\, Q\varphi(Y_i) \\ &= \sum_i \inf\left\{\int_{\Omega_i} \varphi(\nabla w_i)\, dx : w_i \in W^{1,\infty}(\Omega_i),\right. \\ &\qquad\qquad \left. w_i - u_{Y_i} \in W_0^{1,\infty}(\Omega_i)\right\}. \end{aligned}$$

It is easy to ascertain by a patching procedure that the last sum of infima is actually equal to

$$\inf\left\{\int_\Omega \varphi(\nabla w)\,dx : w \in W^{1,\infty}(\Omega), w-u_Y \in W_0^{1,\infty}(\Omega),\right.$$
$$\left. w-u\in W_0^{1,\infty}(\Omega_i),\ \text{for all } i\right\}.$$

Finally,

$$\begin{aligned}\int_\Omega Q\varphi(\nabla u)\,dx =&\inf\left\{\int_\Omega \varphi(\nabla w)\,dx : w \in W^{1,\infty}(\Omega), w-u_Y \in W_0^{1,\infty}(\Omega),\right.\\ &\left. w-u\in W_0^{1,\infty}(\Omega_i),\ \text{for all } i\right\}\\ \geq&\inf\left\{\int_\Omega \varphi(\nabla w)\,dx : w \in W^{1,\infty}(\Omega),\right.\\ &\left. w-u_Y\in W_0^{1,\infty}(\Omega)\right\}\\ =&Q\varphi(Y).\end{aligned}$$

Therefore the inequality (8–8) holds for piecewise affine functions.

Step 2. $Q\varphi$ is upper-semicontinuous.

Let Y be any matrix and Y_j a sequence converging to Y. Let u be any $W^{1,\infty}(\Omega)$ function such that $u-u_Y\in W_0^{1,\infty}(\Omega)$. Consider $u_j(x)=u(x)+(Y_j-Y)x$. Then $u_j-u_{Y_j}\in W_0^{1,\infty}(\Omega)$ and therefore

$$Q\varphi(Y_j)\leq \frac{1}{|\Omega|}\int_\Omega \varphi(\nabla u_j)\,dx = \frac{1}{|\Omega|}\int_\Omega \varphi(\nabla u + Y_j - Y)\,dx.$$

By the upper semicontinuity of φ

$$\limsup_{j\to\infty}\int_\Omega \varphi(\nabla u_j)\,dx \leq \int_\Omega \varphi(\nabla u)\,dx,$$

and

$$\limsup_{j\to\infty} Q\varphi(Y_j)\leq \frac{1}{|\Omega|}\int_\Omega \varphi(\nabla u)\,dx.$$

Since u is arbitrary

$$\limsup_{j\to\infty} Q\varphi(Y_j)\leq Q\varphi(Y).$$

Step 3. Conclusion.

Given any $u\in W^{1,\infty}(\Omega)$ such that $u-u_Y\in W_0^{1,\infty}(\Omega)$ there exists, by Theorem 8.20 (see the Appendix, Section 8), a sequence of piecewise affine functions, $\{u_j\}$, such that

$$\begin{gathered} u_j-u_Y\in W_0^{1,\infty}(\Omega),\\ u_j\to u \quad \text{in } W^{1,p}(\Omega), 1\leq p<\infty,\\ \|\nabla u_j\|_{L^\infty(\Omega)}\leq K\,\|\nabla u\|_{L^\infty(\Omega)}.\end{gathered}$$

Using these estimates on ∇u_j we obtain for some $M > 0$

$$Q\varphi(\nabla u_j) \le M, \quad Q\varphi(\nabla u) \le M,$$

uniformly in Ω. By Fatou's lemma and step 2

$$\int_\Omega (M - Q\varphi(\nabla u))\, dx \le \liminf_{j\to\infty} \int_\Omega (M - Q\varphi(\nabla u_j))\, dx.$$

We conclude by step 1

$$\begin{aligned}\int_\Omega Q\varphi(\nabla u)\, dx &\ge \limsup_{j\to\infty} \int_\Omega Q\varphi(\nabla u_j)\, dx \\ &\ge Q\varphi(Y),\end{aligned}$$

so that $Q\varphi$ is quasiconvex. ■

The next fact refers to the continuity of quasiconvex and rank-one convex functions. It is enough to show that rank-one convex functions are continuous.

Lemma 8.11 *Every finite rank-one convex function is continuous.*

Proof. Observe that given any two matrices A and B, we can find at most $n = \min(m, N)$ matrices A_i, $i = 1, \ldots, n$, such that $A_1 = A$, $A_n = B$ and $\operatorname{rank}(A_i - A_{i-1}) \le 1$ for $i = 2, \ldots, n$. In fact, we can change one full column or row at a time depending on whether $n = N$ or $n = m$.

Let A be fixed and assume that $A^j \to A$. It is clear that $A_i^j \to A$ for all i, and in particular $A_i^j - A_{i-1}^j \to 0$ for all i. Notice that $A_n^j = A$ and $A_1^j = A^j$ for all j. Since φ restricted to the lines determined by A_i^j and A_{i-1}^j is convex, this restriction is continuous as is already well known. Thus

$$\begin{aligned}\lim_{j\to\infty} \varphi(A^j) &= \lim_{j\to\infty} \varphi(A_1^j) = \lim_{j\to\infty} \varphi(A_2^j) \\ &= \ldots \\ &= \lim_{j\to\infty} \varphi(A_{n-1}^j) = \lim_{j\to\infty} \varphi(A_n^j) \\ &= \varphi(A).\end{aligned}$$

■

The last property is a different description of the quasiconvexification of a function that is convenient in some regards.

Lemma 8.12 *Let $\varphi : \mathbf{M}^{m\times N} \to \mathbf{R}^*$ be an upper-semicontinuous function. The following identity holds*

$$Q\varphi = \sup\{\psi : \psi \le \varphi, \psi \text{ quasiconvex}\}.$$

Proof. Let $\tilde{Q}\varphi$ denote the sup in the right-hand side. Since by Lemma 8.10 $Q\varphi$ is an admissible function for such supremum, $Q\varphi \leq \tilde{Q}\varphi$. Conversely, if ψ is any admissible function in the supremum, for any $u \in W_0^{1,\infty}(\Omega)$

$$\psi(Y) \leq \frac{1}{|\Omega|} \int_\Omega \psi(Y + \nabla u)\, dx \leq \frac{1}{|\Omega|} \int_\Omega \varphi(Y + \nabla u)\, dx.$$

Taking the infimum in u, we get $\psi(Y) \leq Q\varphi(Y)$. The arbitrariness of ψ leads to $\tilde{Q}\varphi \leq Q\varphi$. ■

5. $\mathbf{W}^{1,p}$-quasiconvexity

In general, for finite p we say that a function φ defined on matrices is $W^{1,p}$-quasiconvex if for every $Y \in \mathbf{M}^{m\times N}$

$$\varphi(Y) \leq \frac{1}{|\Omega|} \int_\Omega \varphi(\nabla u(x))\, dx,$$

for every $u \in W^{1,p}(\Omega)$ such that $u - u_Y \in W_0^{1,p}(\Omega)$ and any regular domain Ω. Similarly the $W^{1,p}$-quasiconvexification of φ is defined via the infimum

$$Q^p\varphi(Y) = \frac{1}{|\Omega|} \inf \left\{ \int_\Omega \varphi(\nabla u(x))\, dx : u \in W^{1,p}(\Omega), u - u_Y \in W_0^{1,p}(\Omega) \right\}.$$

Clearly $\varphi^{**} \leq Q^1\varphi \leq Q^p\varphi \leq Q^q\varphi \leq Q\varphi$ for $1 \leq p \leq q$. Notice that $Q\varphi$ is in reality $Q^\infty\varphi$.

The following fact relates the property of $W^{1,p}$-quasiconvexity to quasiconvexity. In this context quasiconvexity is $W^{1,\infty}$-quasiconvexity.

Lemma 8.13 *Let φ be a continuous function defined on matrices such that*

$$c \leq \varphi(A) \leq C(1 + |A|^p), \quad C > 0, p \geq 1. \tag{8–9}$$

φ is $W^{1,p}$-quasiconvex if and only if φ is quasiconvex.

Proof. We have to show that if φ is quasiconvex, it is $W^{1,p}$-quasiconvex. Let Y be any matrix and $u \in W^{1,p}(\Omega)$ such that $u - u_Y \in W_0^{1,p}(\Omega)$. Let $u_j \in W^{1,\infty}(\Omega)$ be a sequence converging to u strongly in $W^{1,p}(\Omega)$ and $u_j - u_Y \in W_0^{1,\infty}(\Omega)$. By (8–9)

$$C\left(|\nabla u_j|^p + 1\right) - \varphi(\nabla u_j) \geq 0,$$

and by Fatou's lemma

$$\int_\Omega \left(C\,|\nabla u|^p - \varphi(\nabla u)\right)\, dx \leq \liminf_{j\to\infty} \int_\Omega \left(C\,|\nabla u_j|^p - \varphi(\nabla u_j)\right)\, dx.$$

This implies that

$$\limsup_{j\to\infty} \int_\Omega \varphi(\nabla u_j)\, dx \leq \int_\Omega \varphi(\nabla u)\, dx,$$

and since φ is quasiconvex

$$\varphi(Y) \leq \int_\Omega \varphi(\nabla u)\, dx,$$

so that φ is $W^{1,p}$-quasiconvex. ■

In particular, under the bounds (8–9) the $W^{1,p}$-quasiconvexification of a continuous function is itself $W^{1,p}$-quasiconvex (indeed quasiconvex according to Lemma 8.10) and continuous. Under no growth assumptions, there are still many open, delicate issues concerning $W^{1,p}$-quasiconvexity.

6. The vector case: proof of necessity

We would like to understand the conditions that a family of probability measures should satisfy when coming from a sequence of gradients uniformly bounded in $L^p(\Omega)$. There are three constraints that turn out to be sufficient as well: a spatial compatibility condition, a fundamental property in terms of Jensen's inequality for each individual member and a technical hypothesis. These are i), ii) and iii) below, respectively.

Theorem 8.14 *Let $\{u_j\}$ be a bounded sequence in $W^{1,p}(\Omega)$, $p > 1$, and let $\nu = \{\nu_x\}_{x\in\Omega}$ be the parametrized measure associated to the sequence of gradients $\{\nabla u_j\}$. Then*

i) $\nabla u(x) = \int_{\mathbf{M}^{m\times N}} A\, d\nu_x(A)$, for a.e. $x \in \Omega$ where $u \in W^{1,p}(\Omega)$;

ii) $\int_{\mathbf{M}^{m\times N}} \varphi(A)\, d\nu_x(A) \geq \varphi(\nabla u(x))$ for every $\varphi \in \mathcal{E}^p$ quasiconvex and bounded from below and a.e. $x \in \Omega$;

iii) $\int_\Omega \int_{\mathbf{M}^{m\times N}} |A|^p\, d\nu_x(A)\, dx < \infty$.

For $p = \infty$ ii) and iii) should be changed to

ii') $\int_{\mathbf{M}^{m\times N}} \varphi(A)\, d\nu_x(A) \geq \varphi(\nabla u(x))$ for a.e. $x \in \Omega$ and any φ bounded from below and quasiconvex;

iii') $\operatorname{supp} \nu_x \subset K$ for a.e. $x \in \Omega$ where $K \subset \mathbf{M}^{m\times N}$ is some fixed bounded set.

Assertions i) and iii) are quite simple. In fact, i) is just the representation formula in terms of parametrized measures of the first moment of ν_x. Concerning iii), observe that by Theorem 6.11

$$\int_\Omega \int_{\mathbf{M}^{m\times N}} |A|^p\, d\nu_x(A)\, dx \leq \liminf_{j\to\infty} \int_\Omega \varphi(\nabla u_j)\, dx < \infty.$$

For $p = \infty$, iii') is even easier to prove since the sequence $\{u_j\}$ is uniformly bounded in $W^{1,\infty}(\Omega)$ and therefore the corresponding parametrized measure must be compactly supported uniformly in $x \in \Omega$.

In order to prove ii), let us take any $x \in \Omega$ fixed and write for the moment $\nu = \nu_x$, $Y = \nabla u(x)$. By the localization principle, Theorem 8.4, for a.e. such $x \in \Omega$, there exists a sequence, $\{v_j\}$, converging weakly in $W^{1,p}(\Omega)$ to u_Y and such that $v_j - u_Y \in W_0^{1,p}(\Omega)$ for all j. Moreover, the parametrized measure corresponding to $\{\nabla v_j\}$ is ν, homogeneous. Again the case $p = \infty$ is especially easy to deal with. Assume that $\{v_j\}$ is bounded in $W^{1,\infty}(\Omega)$. If φ is quasiconvex (and thus continuous), since $\{\nabla v_j\}$ is uniformly bounded in $L^\infty(\Omega)$, the sequence (or some suitable subsequence) $\{\varphi(\nabla v_j)\}$ is weakly convergent in $L^1(\Omega)$. Therefore

$$\lim_{j\to\infty}\int_\Omega \varphi(\nabla v_j)\,dx = |\Omega|\int_{\mathbf{M}^{m\times N}}\varphi(A)\,d\nu(A).$$

On the other hand, φ being quasiconvex,

$$|\Omega|\,\varphi(Y)\le\int_\Omega\varphi(\nabla v_j)\,dx,$$

for all j. Hence we obtain

$$\varphi(Y)\le\int_{\mathbf{M}^{m\times N}}\varphi(A)\,d\nu(A),$$

which is ii'). The real problem with finite p lies in the lack of weak compactness for the sequence $\{\varphi(\nabla v_j)\}$. The following remarkable lemma shows how to overcome this fundamental difficulty. We turn back to the case where $\{v_j\}$ converges weakly to u_Y in $W^{1,p}(\Omega)$ for $1<p<\infty$. We refer the reader to the Appendix, Section 8, for some of the notation to be used and some fundamental results about maximal operators to be utilized in the proof of the lemma.

Lemma 8.15 *Let $\{v_j\}$ be a bounded sequence in $W^{1,p}(\Omega)$. There always exists another sequence $\{u_j\}$ of Lipschitz functions ($u_j\in W^{1,\infty}(\Omega)$ for all j) such that $\{|\nabla u_j|^p\}$ is equiintegrable and the two sequences of gradients, $\{\nabla u_j\}$ and $\{\nabla v_j\}$, have the same underlying $W^{1,p}$-parametrized measure.*

Proof. Step 1. Assume furthermore that $v_j\in\mathcal{C}_0^\infty(\mathbf{R}^N)$ and replace Ω by $\mathbf{R}^N$. Consider the sequence $\{\mathcal{M}^*(v_j)\}$ where $\mathcal{M}^*$ is the maximal operator of a function and its gradient. By the remarks recalled in the Appendix, this sequence is bounded in $L^p(\mathbf{R}^N)$. Let $\mu=\{\mu_x\}_{x\in\mathbf{R}^N}$ be the corresponding parametrized measure (possibly for an appropriate subsequence). Consider the truncation operators T_k defined by

$$T_k(s)=\begin{cases} s, & |s|\le k,\\ k\dfrac{s}{|s|}, & k<|s|.\end{cases}$$

Since for fixed k, $\{T_k\mathcal{M}^*(v_j)\}$ is bounded in $L^\infty(\mathbf{R}^N)$,

$$\begin{aligned}\lim_{k\to\infty}\lim_{j\to\infty}\int_{\mathbf{R}^N}|T_k\mathcal{M}^*(v_j)|^p\,dx &= \lim_{k\to\infty}\int_{\mathbf{R}^N}\int_{\mathbf{R}}|T_k(s)|^p\,d\mu_x(s)\,dx\\ &=\int_{\mathbf{R}^N}\int_{\mathbf{R}}|s|^p\,d\mu_x(s)\,dx.\end{aligned}$$

We have used the monotone convergence theorem for the second limit. Notice that

$$g(x) = \int_{\mathbf{R}} |s|^p \, d\mu_x(s)$$

is a $L^1(\mathbf{R}^N)$-function. We can find a subsequence $k(j) \to \infty$ as $j \to \infty$ such that

$$\lim_{j\to\infty} \int_{\mathbf{R}^N} \left|T_{k(j)}\mathcal{M}^*(v_j)\right|^p \, dx = \int_{\mathbf{R}^N} g(x)\, dx. \tag{8–10}$$

On the other hand, by the observation about these truncation operators made after the proof of Lemma 6.3, the parametrized measure associated to the sequence $\left\{T_{k(j)}\mathcal{M}^*(\nabla v_j)\right\}$ is also μ. By Corollary 6.10, we conclude that

$$\left|T_{k(j)}\mathcal{M}^*(v_j)\right|^p \rightharpoonup g \quad \text{in } L^1(\mathbf{R}^N).$$

Let

$$A_j = \{\mathcal{M}^*(v_j) > k(j)\}.$$

Then $|A_j| \to 0$ because $\{\mathcal{M}^*(v_j)\}$ is bounded in $L^p(\mathbf{R}^N)$ and $k(j) \to \infty$. By Lemma 8.21, there exist Lipschitz functions u_j such that $u_j = v_j$ (and therefore $\nabla u_j = \nabla v_j$) outside of A_j and, moreover,

$$|\nabla u_j| \le C(N) k(j), \quad \text{for all } j.$$

The fact that $|A_j| \to 0$ implies that the parametrized measure for both sequences is the same (Lemma 6.3). It follows easily ($\mathcal{M}^*(v_j) \ge |\nabla v_j|$) that

$$|\nabla u_j|^p \le C(N)^p \left|T_{k(j)}\mathcal{M}^*(v_j)\right|^p.$$

Since the right-hand side is equiintegrable in $L^1(\mathbf{R}^N)$ the conclusion of the lemma follows.

Step 2. Approximation. We can assume that $v_j \rightharpoonup u$ in $W^{1,p}(\Omega)$ for some $u \in W^{1,p}(\Omega)$. Moreover, by Lemma 8.3, we can assume that $v_j - u \in W_0^{1,p}(\Omega)$. Let $w_j = v_j - u$ extended by 0 to all of $\mathbf{R}^N$. By density, we can find $z_j \in \mathcal{C}_0^\infty(\mathbf{R}^N)$ such that

$$\|z_j - w_j\|_{W^{1,p}(\mathbf{R}^N)} \to 0, \quad j \to \infty.$$

Apply Step 1 to $\{z_j\}$ and find a sequence of Lipschitz functions, $\{\overline{u}_j\}$, such that $\{|\nabla \overline{u}_j|^p\}$ is equiintegrable in $L^1(\mathbf{R}^N)$ and $|\{\nabla z_j \ne \nabla \overline{u}_j\}| \to 0$. Therefore, again by Lemma 6.3, the parametrized measure for the sequences (considered now restricted to Ω) $\{\nabla \overline{u}_j\}$, $\{\nabla z_j\}$ and $\{\nabla w_j\}$ is the same. Take $u_j = \overline{u}_j|_\Omega + u$. The sequence $\{u_j\}$ verifies the conclusion of the theorem (see Step 2 of the proof of Theorem 8.16). ∎

For proving ii) in Theorem 8.14 in the case p finite, take a bounded sequence in $W^{1,p}(\Omega)$, $\{u_j\}$, generating $\nu = \nu_x$ for fixed $x \in \Omega$. By the lemma just proved, and using Lemma 8.3, we may assume that $u_j - u_Y \in W_0^{1,p}(\Omega)$ where $Y = \nabla u(x)$ and $\{|\nabla u_j|^p\}$ is equiintegrable. In this case, if $\varphi \in \mathcal{E}^p$ is quasiconvex, it is in particular $W^{1,p}$-quasiconvex and

$$|\Omega|\,\varphi(Y) \leq \lim_{j\to\infty} \int_\Omega \varphi(\nabla u_j)\,dx = \int_{\mathbf{M}^{m\times N}} \varphi(A)\,d\nu(A).$$

This ends the proof of Theorem 8.14.

Theorem 8.14 is valid for $p = 1$ if we assume explicitly $u_j \rightharpoonup u$ in $W^{1,1}(\Omega)$.There are however a few steps in the proof that need to be fixed. We do not pursue this direction here.

7. The vector case: proof of sufficiency

This section is devoted to the proof of the result concerning the sufficiency part of Theorem 8.14. In this form it is also valid for $p = 1$.

Theorem 8.16 *Let $\nu = \{\nu_x\}_{x\in\Omega}$ be a family of probability measures supported on $\mathbf{M}^{m\times N}$ such that*

i) $\nabla u(x) = \int_{\mathbf{M}^{m\times N}} A\,d\nu_x(A)$ for some $u \in W^{1,p}(\Omega)$;

ii) $\int_{\mathbf{M}^{m\times N}} \varphi(A)\,d\nu_x(A) \geq \varphi(\nabla u(x))$ for a.e. $x \in \Omega$ and for any $\varphi \in \mathcal{E}^p$ bounded from below and quasiconvex;

iii) $\int_\Omega \int_{\mathbf{M}^{m\times N}} |A|^p\,d\nu_x(A)\,dx < \infty$.

There exist functions $u_j \in W^{1,p}(\Omega)$ such that $\{|\nabla u_j|^p\}$ is weakly convergent in $L^1(\Omega)$ and the parametrized measure associated to $\{\nabla u_j\}$ is ν.

The idea behind the proof is natural. We first take care of the homogeneous case when we do not have any spatial dependence on ν. Property ii) says that if Jensen's inequality holds for all suitable quasiconvex functions then ν can be generated by a sequence of gradients. The inhomogeneity of ν is taken care of by an assembling procedure: we patch together many different individual ν_x.

We begin by treating first the homogeneous case.

Proposition 8.17 *Suppose that $\mu \in (\mathcal{E}^p)'$ is a probability measure for which*

$$Q\varphi(Y) \leq \int_{\mathbf{M}^{m\times N}} \varphi(A)\,d\mu(A), \quad Y = \int_{\mathbf{M}^{m\times N}} A\,d\mu(A), \tag{8–11}$$

whenever $\varphi \in \mathcal{E}^p$. Then μ is a homogeneous $W^{1,p}$-parametrized measure.

Proof. We use the Hahn-Banach theorem. Let T be a linear functional on $(\mathcal{E}^p)'$ in the weak * topology such that $T \geq 0$ on M_Y, a convex set by Lemma 8.5 (the proof of this lemma is also valid for the vector case exactly as it stands). There exists $\psi \in \mathcal{E}^p$ such that

$$0 \leq \langle T, \nu \rangle = \langle \psi, \nu \rangle = \int_{\mathbf{M}^{m \times N}} \psi(A)\, d\nu(A), \quad \nu \in M_Y.$$

For $\nu = \overline{\delta_{\nabla u}}$, $u \in W^{1,p}(\Omega)$, $u - u_Y \in W_0^{1,p}(\Omega)$,

$$0 \leq \int_\Omega \psi(\nabla u)\, dx. \tag{8–12}$$

Therefore, $Q\psi(Y) \geq 0$. Thus by (8–11),

$$0 \leq Q\psi(Y) \leq \int_{\mathbf{M}^{m \times N}} \psi(A)\, d\mu(A) = \langle T, \mu \rangle.$$

Therefore, $\mu \in \overline{\operatorname{co}(M_Y)} = \overline{M_Y}$ where closure is meant in the weak * sense. Since $\mathcal{E}^p$ is separable, bounded sets in $(\mathcal{E}^p)'$ endowed with the weak * topology are metrizable, and convergence can be characterized by sequences. Hence in a bounded neighborhood of μ there exists a sequence $\{u^k\} \subset W^{1,p}(\Omega)$, $u_k - u_Y \in W_0^{1,p}(\Omega)$ such that

$$\int_{\mathbf{M}^{m \times N}} \psi(A)\, d\mu(A) = \lim_{k \to \infty} \int_\Omega \psi(\nabla u^k)\, dx \quad \text{for any } \psi \in \mathcal{E}^p. \tag{8–13}$$

Let ν be the $W^{1,p}$-parametrized measure associated to $\{\nabla u^k\}$. By the averaging procedure we may assume ν to be homogeneous (notice that (8–13) does not change in this process). Clearly $\mu = \nu$, since as a consequence of (8–13),

$$\langle \psi, \mu \rangle = \langle \psi, \nu \rangle$$

for all $\psi \in \mathcal{C}_0(\mathbf{M}^{m \times N})$. ■

Theorem 8.18 *A probability measure μ in $(\mathcal{E}^p)'$ is a homogeneous $W^{1,p}$-parametrized measure if*

$$\varphi\left(\int_{\mathbf{M}^{m \times N}} A\, d\mu(A)\right) \leq \int_{\mathbf{M}^{m \times N}} \varphi(A)\, d\mu(A), \tag{8–14}$$

for every $\varphi \in \mathcal{E}^p$ which is quasiconvex and bounded from below.

Proof. Assume $\psi \in \mathcal{E}^p$, and set

$$\psi_n = \max(\psi, \alpha_n) = \psi \chi_{\{\psi \geq \alpha_n\}} + \alpha_n \chi_{\{\psi < \alpha_n\}}$$

where $\alpha_n \to -\infty$. $\psi_n \to \psi$ monotonically and $\psi_n \in \mathcal{E}^p$. The following lemma asserts that $Q\psi_n$ does belong to $\mathcal{E}^p$ as well.

Lemma 8.19 *If $\psi \in \mathcal{E}^p$ and $\psi \geq C$ then $Q\psi \in \mathcal{E}^p$ (and $Q\psi \geq C$).*

Proof. Assume first that $C = 0$. Let

$$\lim_{|A|\to\infty} \frac{\psi(A)}{1+|A|^p} = \alpha \geq 0.$$

If $\alpha = 0$ then trivially

$$\lim_{|A|\to\infty} \frac{Q\psi(A)}{1+|A|^p} = 0$$

as well. Let $\alpha > 0$, and $0 < \epsilon < \alpha$. There exists M_ϵ such that for $|A| \geq M_\epsilon$

$$\psi(A) \geq (\alpha - \epsilon)\,|A|^p + (\alpha - \epsilon), \quad \alpha - \epsilon > 0$$

On the other hand $\psi(A) \geq -C_\epsilon$, $C_\epsilon > 0$, if $|A| \leq M_\epsilon$. Altogether we have for any A

$$\psi(A) \geq (\alpha - \epsilon)\,(|A|^p - M_\epsilon^p) - C_\epsilon.$$

Since the right-hand side is a convex function, we conclude by Lemma 8.12,

$$Q\psi(A) \geq (\alpha - \epsilon)\,(|A|^p - M_\epsilon^p) - C_\epsilon,$$

and taking limits for $|A| \to \infty$ we get

$$\liminf_{|A|\to\infty} \frac{Q\psi(A)}{1+|A|^p} \geq \alpha - \epsilon.$$

The arbitrariness of $\epsilon > 0$ and the fact that $Q\psi \leq \psi$ enables us to write

$$\alpha \leq \liminf_{|A|\to\infty} \frac{Q\psi(A)}{1+|A|^p} \leq \lim_{|A|\to\infty} \frac{\psi(A)}{1+|A|^p} = \alpha.$$

This implies the conclusion of the lemma.

If C is not 0, we apply the preceding arguments to $\tilde{\psi} = \psi - C$. ■

We go back to the proof of Theorem 8.18. By hypothesis, since $Q\psi_n \in \mathcal{E}^p$ is quasiconvex

$$Q\psi(Y) \leq Q\psi_n(Y) \leq \int_{\mathbf{M}^{m\times N}} Q\psi_n(A)\,d\nu(A) \leq \int_{\mathbf{M}^{m\times N}} \psi_n(A)\,d\nu(A).$$

But by monotone convergence

$$\int_{\mathbf{M}^{m\times N}} \psi_n(A)\,d\nu(A) \to \int_{\mathbf{M}^{m\times N}} \psi(A)\,d\nu(A).$$

We now use Proposition 8.17 to conclude. ■

Proof of theorem 8.16.

Step 1. Assume the function $u \in W^{1,p}(\Omega)$ in i) and ii) is 0. It is sufficient to find a sequence of $W^{1,p}(\Omega)$-functions with the property

$$\lim_{k\to\infty}\int_\Omega \xi(x)\varphi(\nabla u^k(x))\,dx = \int_\Omega \xi(x)\int_{\mathbf{M}^{m\times N}}\varphi(A)\,d\nu_x(A)\,dx,$$

for all $\xi \in \Gamma$ and $\varphi \in \mathcal{S}$, where Γ and $\mathcal{S}$ are dense, countable subsets of $L^1(\Omega)$ and $\mathcal{C}_0(\mathbf{M}^{m\times N})$, respectively. The equiintegrability of $\{|\nabla u^k|^p\}$ is guaranteed by Lemma 8.15.

Apply Lemma 7.9 for $p = \infty$ and $q = 1$ to the functions in $\mathcal{S}$ taking $r_k = 1/k$ and find $a_i^k \in \Omega \setminus N$, $\epsilon_i^k \leq 1/k$ with

$$\int_\Omega \xi(x)\overline{\varphi}(x)\,dx = \lim_{k\to\infty}\sum_i \overline{\varphi}(a_i^k)\int_{a_i^k+\epsilon_i^k\Omega}\xi(x)\,dx, \tag{8–15}$$

for all $\xi \in L^1(\Omega)$ and $\varphi \in \mathcal{S}$. We may assume that for $a_i^k \notin N$, $|N| = 0$, by ii) and Theorem 8.18, $\nu_{a_i^k}$ is a homogeneous $W^{1,p}$-parametrized measure so that for $\left\{u_j^{k,i}\right\}_j$,

$$u_j^{k,i} \rightharpoonup 0 \text{ in } W^{1,p}(\Omega)$$

(recall that we are taking $u \equiv 0$) and

$$\lim_{j\to\infty}\int_\Omega \xi(x)\varphi(\nabla u_j^{k,i}(x))\,dx = \overline{\varphi}(a_i^k)\int_\Omega \xi(x)\,dx, \qquad \xi \in L^\infty(\Omega).$$

Moreover, by Lemma 8.3, we may assume that $u_j^{k,i} \in W_0^{1,p}(\Omega)$. Define the sequence u_k by

$$u_k(x) = \epsilon_i^k\, u_j^{k,i}\left(\frac{x-a_i^k}{\epsilon_i^k}\right)$$

if $x \in a_i^k + \epsilon_i^k\Omega$ and $u_k = 0$ otherwise, where $j = j(k,i)$ will be chosen soon.

Let $\Gamma \times \mathcal{S} = \bigcup_k D_k$ with D_k finite and $D_k \subset D_{k+1}$. For k, i fixed we take $j = j(k,i)$ sufficiently large so that for every $(\xi,\varphi) \in D_k$

$$\left|(\epsilon_i^k)^N\int_\Omega \xi(a_i^k+\epsilon_i^k y)\varphi(\nabla u_h^{k,i}(y))\,dy - \int_{a_i^k+\epsilon_i^k\Omega}\xi(x)\,dx\;\overline{\varphi}(a_i^k)\right| \leq \frac{1}{k2^i}.$$

Having in mind these estimates and (8–15) we obtain for any $(\xi,\varphi) \in \Gamma\times\mathcal{S}$

$$\begin{aligned}\lim_{k\to\infty}\int_\Omega \xi(x)\varphi(\nabla u_k(x))\,dx &= \lim_{k\to\infty}\sum_i(\epsilon_i^k)^N\int_\Omega \xi(a_i^k+\epsilon_i^k y)\varphi(\nabla u_h^{k,i}(y))\,dy\\ &= \lim_{k\to\infty}\sum_i\int_{a_i^k+\epsilon_i^k\Omega}\xi(x)\,dx\;\overline{\varphi}(a_i^k)\\ &= \int_\Omega \xi(x)\overline{\varphi}(x)\,dx,\end{aligned}$$

as desired.

Step 2. Given a family $\nu = \{\nu_x\}_{x\in\Omega}$ satisfying i), ii) and iii), consider

$$\langle \tilde{\nu}_x, \psi \rangle = \langle \nu_x, \psi(\cdot - \nabla u(x)) \rangle.$$

It is elementary to check that $\tilde{\nu}$ is in the situation of Step 1 (we can always normalize the quasiconvex functions entering in ii) by requiring $\varphi(0) = 0$). Therefore there exists a sequence $\{v_j\}$ generating $\tilde{\nu}$ as a $W^{1,p}$-parametrized measure and such that $\{|\nabla v_j|^p\}$ is equiintegrable. We claim that $u_j = v_j + u$ generates ν. If this is so we have completed the proof of the theorem. In order to see that claim, let $\psi(x, A)$ be a Carathéodory function and let $\tilde{\psi}(x, A) = \psi(x, A + \nabla u(x))$, itself a Carathéodory function. Hence

$$\begin{aligned}
\lim_{j\to\infty} \int_\Omega \psi(x, \nabla u_j)\, dx &= \lim_{j\to\infty} \int_\Omega \tilde{\psi}(x, \nabla v_j)\, dx \\
&= \int_\Omega \int_{\mathbf{M}^{m\times N}} \tilde{\psi}(x, A)\, d\tilde{\nu}_x(A)\, dx \\
&= \int_\Omega \int_{\mathbf{M}^{m\times N}} \tilde{\psi}(x, A - \nabla u(x))\, d\nu_x(A)\, dx \\
&= \int_\Omega \int_{\mathbf{M}^{m\times N}} \psi(x, A)\, d\nu_x(A)\, dx.
\end{aligned}$$

The arbitrariness of ψ implies that the parametrized measure associated to $\{\nabla u_j\}$ is ν. ■

Theorem 8.16 remains true for the case $p = \infty$ changing conditions ii) and iii) to ii') and iii') as in the previous section. The proof of this involves some further technicalities although the tools are the same as in Theorem 8.16. Since the case $p = \infty$ is not relevant in Chapters 3 and 4, we do not include the proof here.

8. Appendix

1. We recall an approximation result by piecewise affine functions which is very useful in many different settings.

A function $u \in W^{1,\infty}(\Omega)$ is called piecewise affine if Ω can be decomposed in a finite union $\overline{\Omega} = \bigcup_i \overline{\Omega}_i$ and ∇u is constant on each Ω_i.

Theorem 8.20 *Let Ω be a bounded domain with Lipschitz boundary and $u \in W_0^{1,\infty}(\Omega)$. There exist functions $u_j \in W_0^{1,\infty}(\Omega)$, piecewise affine, such that*

$$\begin{aligned}
&u_j \to u \quad \text{in } W^{1,p}(\Omega), 1 \le p < \infty, \\
&\|\nabla u_j\|_{L^\infty(\Omega)} \le K \|\nabla u\|_{L^\infty(\Omega)},
\end{aligned}$$

with K a constant not depending on j.

2. We also state some facts about maximal operators. For any $v \in \mathcal{C}_0^\infty(\mathbf{R}^N)$, we set

$$\mathcal{M}^* v(x) = \mathcal{M}(|v(x)|) + \mathcal{M}(|\nabla v(x)|),$$

where

$$\mathcal{M} f(x) = \sup_{r>0} \frac{1}{|B_r|} \int_{B_r(x)} |f(z)|\, dz$$

is the maximal function of f. It is well known that if $v \in \mathcal{C}_0^\infty(\mathbf{R}^N)$, $\mathcal{M}^* v \in \mathcal{C}(\mathbf{R}^N)$ and

$$\|\mathcal{M}^* v\|_{L^p(\mathbf{R}^N)} \leq C(N,p)\, \|v\|_{W^{1,p}(\mathbf{R}^N)}, \quad 1 < p \leq \infty,$$

and, in particular, for any $\lambda > 0$,

$$|\{\mathcal{M}^* v \geq \lambda\}| \leq C(N,p) \lambda^{-p}\, \|v\|^p_{W^{1,p}(\mathbf{R}^N)}, \quad 1 < p < \infty.$$

This last inequality is also valid for $p = 1$ even though the previous one is not.

Lemma 8.21 *Let $v \in \mathcal{C}_0^\infty(\mathbf{R}^N)$ and $\lambda > 0$. Set $H^\lambda = \{\mathcal{M}^* v < \lambda\}$. Then*

$$\frac{|v(x) - v(y)|}{|x - y|} \leq C(N)\lambda, \quad x, y \in H^\lambda,$$

where $C(N)$ depends only on N.

It is also interesting to remember that any Lipschitz function defined on a subset of $\mathbf{R}^N$ may be extended to all of $\mathbf{R}^N$ without increasing its Lipschitz constant.

9. Bibliographical remarks

The basic ingredients of homogenization and localization in the format developed here are contained in [191]. The basic construction on the proof of Lemma 8.6 has been known for many years.

All the material related to quasiconvexity can also be studied in many different sources. Indeed, a fairly large number of works in the last two decades have been directed towards the understanding of the quasiconvexity condition and the property of weak lower semicontinuity for variational integrals. Once again we do not try to exhaust the bibliography on this subject. Some of these references are [2], [3], [4], [29], [37], [61], [95], [96], [97], [98], [151], [166], [193], [216], [220], [225], [233], [238], [269], [289], [309]. The rank-one convexity condition has also been investigated extensively: [32], [96], [256], [257], [295], [307], [319]. The $W^{1,p}$-quasiconvexity condition has been studied in great detail in [38]. We refer to this paper for examples and further discussion.

The proof of the necessity part for Theorem 8.15 is based on ideas and technical results introduced in [145]. It is somehow a shorter, more direct version than the one contained in [4]. An alternative approach based on extra integrability for minimizing sequences was developed in [226]. The sufficiency part, Theorem 8.16, has been taken from [191]. Recently, another method has been proposed in [200] based on the Hodge decomposition.

The facts in the Appendix are well known. The approximation by piecewise affine functions can be found in [118]. [291] is a standard reference for basic properties of maximal operators. See also [135].

Chapter 9
Quasiconvexity and Rank-one Convexity

1. Introduction

The motivation for this chapter is two-fold. On the one hand, since Jensen's inequality has played a prominent role in our approach to weak lower semicontinuity, our analysis would be somehow incomplete without any reference to this inequality with respect to rank-one convex functions. Because quasiconvexity implies rank-one convexity, probability measures satisfying Jensen's inequality with respect to the class of rank-one convex functions are indeed examples of gradient parametrized measures. It turns out that this family of probability measures can be understood, at least conceptually, in a nice constructive way. They are called *laminates* to emphasize its layering structure. As a matter of fact, laminates are almost the only way to produce explicitly examples of gradient parametrized measures. It is true that the Riemann-Lebesgue lemma allows one to consider gradient parametrized measures associated with periodic gradients. The problem is that we do not know how to decide whether they are laminates or not. The importance of laminates in the description of some equilibrium states for crystals has been stressed in Chapter 5. They are also important in the theory of composite materials and homogenization.

The second goal of this chapter is to show that rank-one convexity does not imply quasiconvexity in general. There is a duality between gradient parametrized measures and quasiconvexity, and laminates and rank-one convexity. Jensen's inequality is the link. In this sense, the problem of deciding if rank-one convexity implies quasiconvexity is equivalent to deciding if every gradient parametrized measure is a laminate. The two problems are equally difficult at first sight. We will show an explicit example of a gradient parametrized measure (a microstructure) constructed through the Riemann-Lebesgue lemma from a periodic gradient which is not a laminate. At the same time, we will find an explicit example of a rank-one convex function that is not quasiconvex. Both examples are intimately connected. They are valid only if the dimension of the target space for deformations is three or more. For dimension two the problem is still open. We will consider a direct extension of the counterexample to dimension two and show how it dramatically fails. New ideas are needed to clarify the two-dimensional situation.

An equivalent way of understanding rank-one convexity that is helpful sometimes is the following. A function Ψ defined on matrices is rank-one convex if and only if the functions of one real variable $g(t) = \Psi(Y + tF)$ are convex for all matrices Y and F with $\operatorname{rank}(F) = 1$. If Ψ is smooth, this is the case if

$$\frac{d^2}{dt^2}\Psi(Y + tF)\bigg|_{t=0} \geq 0$$

for all such Y and F. If this second derivative is written in terms of the derivatives of Ψ we obtain the Legendre-Hadamard condition for rank-one convexity

$$\sum_{i,j,k,l} \frac{\partial^2 \Psi}{\partial x_{ij} \partial x_{kl}}(Y) a_i a_k n_j n_l \geq 0$$

for all matrices $Y \in \mathbf{M}^{m\times N}$ and vectors $a \in \mathbf{R}^m$ and $n \in \mathbf{R}^N$. We have taken here $F = a \otimes n$.

2. Laminates

Let us briefly recall (in a slightly different form) how the rank-one convexity condition was introduced in Chapter 1. Let $Y_i \in \mathbf{M}^{m\times N}$, $i = 1, 2$, $a \in \mathbf{R}^m$ and a unit vector $n \in \mathbf{R}^N$ be given in such a way that

$$Y_1 - Y_2 = a \otimes n. \tag{9–1}$$

If χ_t is the characteristic function of the interval $(0, t)$ in $(0, 1)$ extended by periodicity, the parametrized measure associated to the sequence of gradients

$$\nabla u_j(x) = Y_2 + \chi_t(jx \cdot n) a \otimes n,$$
$$u_j(x) = Y_2 x + \int_0^{jx\cdot n} \chi_t(s)\, ds\; a,$$

is

$$\nu_x = \nu = t\delta_{Y_1} + (1-t)\delta_{Y_2}, \quad x \in \Omega, \tag{9–2}$$

where Ω is any bounded domain in $\mathbf{R}^N$. Therefore the probability measure ν in (9–2) is a gradient parametrized measure (by this we actually mean $W^{1,\infty}$-parametrized measure) for any $t \in [0, 1]$ provided the compatibility condition (9–1) holds. Furthermore, by Lemma 8.3, we can assume without loss of generality that $u_j - u_Y \in W_0^{1,\infty}(\Omega)$, $Y = tY_1 + (1-t)Y_2$. In this case ∇u_j takes on the values Y_1 and Y_2 except in small sets E_j, $|E_j| \to 0$. We would like to go one step further as described in the proof of Lemma 8.6. Assume, in addition to (9–1), that

$$\begin{aligned} Y_2 &= t_0 Y_2^{(1)} + (1-t_0) Y_2^{(2)}, \quad t_0 \in (0,1), \\ Y_2^{(1)} - Y_2^{(2)} &= b \otimes e, \end{aligned} \tag{9–3}$$

where $b \in \mathbf{R}^m$ and $e \in \mathbf{R}^N$ is another unit vector. Let Ω_i^j be the part of Ω where $\nabla u_j = Y_i$. For j and i fixed, based on the compatibility condition between $Y_2^{(1)}$ and $Y_2^{(2)}$, we can construct a sequence of gradients $\left\{\nabla v_k^{ji}\right\}$, $v_k^{ji} - u_{Y_2} \in W_0^{1,\infty}(\Omega_i^j)$, whose values essentially alternate between $Y_2^{(1)}$ and $Y_2^{(2)}$ with preassigned frequency $t_0 \in (0,1)$ and normal e to the layers. Let E_k^{ji} be the set where ∇v_k^{ji} does not take either of the two values $Y_2^{(1)}$ or $Y_2^{(2)}$. Choose $k = k(j,i)$ in such a way that

$$\frac{\left|E_k^{ji}\right|}{\left|\Omega_i^j\right|} \to 0$$

as $j \to \infty$ uniformly in $i = 1, 2$. Define

$$u^{(j)}(x) = \begin{cases} v_{k(j,i)}^{ji}(x), & x \in \Omega_i^j, \\ u_j(x), & \text{else.} \end{cases}$$

This sequence $\left\{u^{(j)}\right\}$ is uniformly bounded in $W^{1,\infty}(\Omega)$ and satisfies $u^{(j)} - u_Y \in W_0^{1,\infty}(\Omega)$. The parametrized measure associated to $\left\{\nabla u^{(j)}\right\}$ is

$$\nu = t\delta_{Y_1} + (1-t)\left(t_0 \delta_{Y_2^{(1)}} + (1-t_0)\delta_{Y_2^{(2)}}\right), \tag{9–4}$$

homogeneous. The probability measure in (9–4) is a gradient parametrized measure provided we have the compatibility conditions (9–1) and (9–3).

It is not difficult to generalize this construction when a finite number of matrices is involved if we have the rank-one condition in a recursive way. This basic construction has been referred to as "layers within layers" in the literature and reflects accurately the situation. It motivates the following definition.

Definition 9.1 *A set of pairs $\{(t_i, Y_i)\}_{1 \le i \le l}$ where $t_i > 0$, $\sum_i t_i = 1$, $Y_i \in \mathbf{M}^{m \times N}$ is said to satisfy the (H_l) condition if:*

i) for $l = 2$, rank$(Y_1 - Y_2) \le 1$;

ii) for $l > 2$ and possibly after a permutation of indices, rank$(Y_1 - Y_2) \le 1$ and if we set

$$\begin{aligned} s_1 &= t_1 + t_2, \quad Z_1 = \frac{t_1}{s_1} Y_1 + \frac{t_2}{s_1} Y_2, \\ s_i &= t_{i+1}, \quad Z_i = Y_{i+1}, \quad 2 \le i \le l-1, \end{aligned}$$

the set of pairs $\{(s_i, Z_i)\}_{1 \le i \le l-1}$ satisfies the (H_{l-1}) condition.

An immediate consequence of our previous discussion is

Proposition 9.2 *If $\{(t_i, Y_i)\}_{1\le i\le l}$ satisfies the (H_l) condition the probability measure $\nu = \sum_i t_i\delta_{Y_i}$ is a gradient parametrized measure.*

We can even take weak * limits in the sense of measures for sequences of finite convex combinations of Dirac masses verifying (H_l) conditions. These weak limits will also be gradient parametrized measures: the argument is elementary and involves taking diagonal sequences. Notice that in fact the set of homogeneous gradient parametrized measures is weak * closed. This remark motivates the definition of laminate. In a sense, these weak * limits correspond to (H_l) conditions when $l \to \infty$ and in this case we talk about infinite order laminates.

Definition 9.3 *Let ν be a probability measure supported on $\mathbf{M}^{m\times N}$ and let $K = supp(\nu)$ be a compact set. ν is a laminate if there exists a sequence of sets of pairs $\{(t_i^k, Y_i^k)\}_{1\le i\le k}$, $(k \ge 2)$, verifying the (H_k) condition such that $\sum_i t_i^k \delta_{Y_i^k} \overset{*}{\rightharpoonup} \nu$ in the sense of measures.*

Proposition 9.4 *Every laminate is a homogeneous gradient parametrized measure.*

There are indeed two main ways of producing explicitly gradient parametrized measures. One relies on laminates and the other on the Riemann-Lebesgue lemma for periodic gradients. The difficulty is in deciding whether or not we get the same measures with both. One of the main goals of this chapter is to show that there are gradient parametrized measures coming from periodic gradients (microstructures) that **are not** laminates.

It may be extremely hard to decide if a given probability measure is a laminate even for innocent looking examples. For instance, consider the matrices

$$A_1 = \begin{pmatrix} -3 & 0 \\ 0 & -1 \end{pmatrix}, \quad A_2 = \begin{pmatrix} 1 & 0 \\ 0 & -3 \end{pmatrix}$$
$$A_3 = \begin{pmatrix} 3 & 0 \\ 0 & 1 \end{pmatrix}, \quad A_4 = \begin{pmatrix} -1 & 0 \\ 0 & 3 \end{pmatrix}.$$

We claim that the probability measure

$$\nu = \frac{8}{15}\delta_{A_1} + \frac{4}{15}\delta_{A_2} + \frac{2}{15}\delta_{A_3} + \frac{1}{15}\delta_{A_4}$$

is a laminate. The reader is invited to find a sequence of (H_l) conditions whose weak * limit is ν. Consider the matrices

$$B_1 = \begin{pmatrix} 1 & 0 \\ 0 & -1 \end{pmatrix}, \quad B_2 = \begin{pmatrix} 1 & 0 \\ 0 & 1 \end{pmatrix}$$
$$B_3 = \begin{pmatrix} -1 & 0 \\ 0 & 1 \end{pmatrix}, \quad B_4 = \begin{pmatrix} -1 & 0 \\ 0 & -1 \end{pmatrix}.$$

It is easy to check that

$$\begin{aligned} B_4 &= \frac{1}{2}A_1 + \frac{1}{2}B_1, \quad \operatorname{rank}(A_1 - B_1) = 1,\\ B_1 &= \frac{1}{2}A_2 + \frac{1}{2}B_2, \quad \operatorname{rank}(A_2 - B_2) = 1,\\ B_2 &= \frac{1}{2}A_3 + \frac{1}{2}B_3, \quad \operatorname{rank}(A_3 - B_3) = 1,\\ B_3 &= \frac{1}{2}A_4 + \frac{1}{2}B_4, \quad \operatorname{rank}(A_4 - B_4) = 1. \end{aligned}$$

If we use these identities one after the other in cyclic order and add the resulting series for the coefficients we find after an infinite number of steps that ν is truly a laminate.

A necessary condition for a probability measure to be a laminate is Jensen's inequality for rank-one convex functions: if $\varphi : \mathbf{M}^{m\times N} \to \mathbf{R}$ is rank-one convex and ν is a laminate then

$$\varphi\left(\int_K A\, d\nu(A)\right) \leq \int_K \varphi(A)\, d\nu(A). \tag{9–5}$$

This is easy to derive because of the recursive way in which (H_l) conditions are defined. It turns out that this condition is also sufficient. The proof of this is the aim of the next section. The connection between laminates and rank-one convexity, although intuitively clear, is made precise by the following definition and lemma. Given $\varphi : \mathbf{M}^{m\times N} \to \mathbf{R}^*$, its rank-one convexification is defined by

$$R\varphi(Y) = \inf\left\{\sum_i t_i\varphi(Y_i) : Y = \sum_i t_iY_i, \{(t_i, Y_i)\}_{1\leq i\leq l} \text{ verifies } (H_l)\right\}, \tag{9–6}$$

or equivalently by

$$R\varphi(Y) = \inf\left\{\int_{\mathbf{M}^{m\times N}} \varphi(A)d\nu(A) : \nu \text{ is a laminate and } Y = \int_{\mathbf{M}^{m\times N}} A d\nu(A)\right\}.$$

Lemma 9.5 *1. If $\{(t_i, Y_i)\}_{1\leq i\leq l}$ and $\{(s_j, Z_j)\}_{1\leq j\leq k}$ satisfy the (H_l) and (H_k) conditions, respectively, then*

$$\left\{\{(tt_i, Y_i)\}_{1\leq i\leq l}, \{((1-t)s_j, Z_j)\}_{1\leq j\leq k}\right\}$$

verifies the (H_{l+k}) condition for any $t \in [0,1]$ provided that

$$\operatorname{rank}\left(\sum_i t_iY_i - \sum_j s_jZ_j\right) \leq 1.$$

2. If $\varphi : \mathbf{M}^{m\times N} \to \mathbf{R}^*$ is bounded from below, $R\varphi$ is a rank-one convex function such that

$$R\varphi = \sup\left\{\psi \leq \varphi : \psi \text{ is rank-one convex}\right\}.$$

The proof of this lemma is an interesting exercise of manipulation of inequalities. It is elementary and left to the reader.

3. Characterization of laminates

We want to establish the characterization of laminates in terms of Jensen's inequality for rank-one convex functions. There are two preliminary steps interesting in their own right which we prove below. The first is an easy-to-prove version of this characterization.

Theorem 9.6 *Let K be a compact set of matrices. A probability measure, ν, supported in K is a laminate if and only if there exists a closed ball B containing K such that*

$$R(\hat{\varphi})(Y) \leq \int_K \varphi(A)\, d\nu(A), \quad Y = \int_K A\, d\nu(A), \tag{9–7}$$

where

$$\hat{\varphi} = \begin{cases} \varphi, & \text{in } B, \\ +\infty, & \text{in } \mathbf{M}^{m\times N} \setminus B. \end{cases}$$

Proof. Suppose there exists a closed ball B containing K with the property indicated, and let $Y \in B$ be the first moment of ν. Consider the set of probability measures

$$M_Y = \Big\{ \sum_i t_i \delta_{Y_i} : t_i > 0, \sum_i t_i = 1, Y_i \in B, Y = \sum_i t_i Y_i, \\ \{(t_i, Y_i)\} \text{ satisfies } (H_l) \Big\}.$$

We claim

i) M_Y is convex: this is easy after Lemma 9.5 and the convexity of B;

ii) $\nu \in \overline{M_Y}$ in the weak * topology of measures.

For ii), we use once again the Hahn-Banach theorem. Let T be a linear, continuous functional with respect to the weak * topology on the space of measures supported in B, so that T can be represented by a continuous function, φ, in B. We can certainly extend φ to all of space in a continuous fashion so that we may regard φ as continuous in all of space. Suppose that

$$\langle \varphi, \mu \rangle = \langle T, \mu \rangle \geq 0,$$

for all $\mu = \sum_i t_i \delta_{Y_i} \in M_Y$. By (9–6),

$$R(\hat{\varphi})(Y) \geq 0,$$

and

$$\int_K \varphi(A)\, d\nu(A) = \langle \varphi, \nu \rangle \geq 0,$$

if (9–7) holds. Therefore $\nu \in \overline{\operatorname{co}(M_Y)} = \overline{M_Y}$ and this means that ν is a laminate according to Definition 9.3.

The converse is elementary since the weak * convergence

$$\sum_i t_i^k \delta_{Y_i^k} \overset{*}{\rightharpoonup} \nu$$

implies that the set of matrices $\{Y_i^k\}$ is bounded if $\operatorname{supp}(\nu)$ is compact. Otherwise, if $Y_i^k \to \infty$ for some sequence we may take a continuous function, φ, such that $\varphi \geq 0$, $\varphi|_K = 0$ and $\varphi(Y_i^k) = 1/t_i^k$. This contradicts the weak * convergence. (9–7) follows from (9–6). ■

The proof of this theorem has been modeled after the experience we have using the Hahn-Banach theorem. We are after a characterization of laminates in terms of Jensen's inequality for rank-one convex functions that are finite everywhere. The following lemma will be crucial in building the bridge from Theorem 9.6 to Theorem 9.8 below.

Lemma 9.7 *Let $\varphi : \mathbf{M}^{m\times N} \to \mathbf{R}$ be a continuous function, bounded from below, such that $\varphi^{**} = \varphi$ outside some closed ball B. If $Y \in B$ then*

$$R\varphi(Y) = \inf\left\{\sum_i t_i\varphi(Y_i) : Y = \sum_i t_i Y_i, Y_i \in B,\right.$$
$$\left.\{(t_i, Y_i)\}_{1\leq i\leq l} \text{ verifies } (H_l)\right\}.$$

φ^{**} denotes the convexification of φ. The condition $\varphi^{**} = \varphi$ can be replaced by $R\varphi = \varphi$ outside B and the conclusion still holds.

Proof. Consider the function

$$\psi = \begin{cases} R(\hat{\varphi}), & \text{in } B, \\ \varphi, & \text{in } \mathbf{M}^{m\times N} \setminus B. \end{cases}$$

We claim that ψ is rank-one convex. The proof of this relies on the same fact used in dimension one. Namely, if g is a continuous function defined on $\mathbf{R}$ such that $g = g^{**}$ outside some closed interval J, then for $\alpha \in J$

$$g^{**}(\alpha) = \inf\{t_1 g(\alpha_1) + t_2 g(\alpha_2) : t_1\alpha_1 + t_2\alpha_2 = \alpha, t_1 + t_2 = 1, \alpha_i \in J\}.$$

This is left as an exercise to the reader. Notice that $g = g^{**}$ even in the endpoints of J.

Observe that $\varphi^{**} \leq \psi$ (this is easy to establish) implies

$$\varphi = \varphi^{**} \leq \psi^{**} \leq \psi = \varphi$$

outside B. Therefore by applying the result in dimension one to each function $g(t) = \psi(Y + tF)$ for Y and F fixed matrices and $\operatorname{rank}(F) = 1$ we prove the claim.

By Lemma 9.5,

$$\psi \leq R(\varphi) \leq R(\hat{\varphi}) = \psi$$

in B, and the fact $R(\varphi) = R(\hat{\varphi})$ implies the conclusion of the lemma. ■

Theorem 9.8 *Let ν be a probability measure compactly supported in $\mathbf{M}^{m\times N}$, and let $K = supp(\nu)$. ν is a laminate if and only if*

$$\varphi\left(\int_K A\,d\nu(A)\right) \le \int_K \varphi(A)\,d\nu(A),$$

for all rank-one convex functions $\varphi : \mathbf{M}^{m\times N} \to \mathbf{R}$.

Proof. Let $Y \in \mathrm{co}(K)$ be the first moment of ν. Consider two closed balls, $B_{1/2}$ and B_1, of radii $r_1/2$ and r_1, respectively, and such that $K \subset B_{1/2}$. Assume them to be centered at the origin. Let φ be any continuous function defined over matrices and finite everywhere. Set

$$M = \max_{B_1} \varphi, \quad g(r) = \min_{|A|=r} \varphi(A).$$

Because g is bounded in $[0, r_1/2]$, choose $c \in \mathbf{R}$ such that for $f(r) = c(r-r_1)+M$ we have $f(r) < g(r)$ for $r \in [0, r_1/2]$, and define

$$\psi(A) = \begin{cases} \varphi(A), & A \in B_{1/2}, \\ \max\{\varphi(A), f(|A|)\}, & A \in B_1 \setminus B_{1/2}, \\ f(|A|), & A \in \mathbf{M}^{m\times N} \setminus B_1. \end{cases}$$

Clearly ψ is continuous and finite everywhere and $f(|A|) \le \psi(A)$, which in turn implies, since $f(|A|)$ is convex and $\psi(A) = f(|A|)$ outside B_1, that $\psi^{**} = \psi$ outside B_1. By Lemma 9.5, $R\psi$ is rank-one convex and finite (notice that ψ is bounded from below) so that by hypothesis

$$R\psi(Y) \le \int_K \psi(A)\,d\nu(A) = \int_K \varphi(A)\,d\nu(A),$$

because $\psi = \varphi$ in $K \subset B_{1/2}$. We claim that $R(\hat{\varphi})(Y) \le R\psi(Y)$ where as before $\hat{\varphi}$ is defined as $+\infty$ outside B_1. If this is so, then by Theorem 9.6 the proof is finished.

By Lemma 9.7, in order to find the rank-one convexification of ψ it is enough to restrict attention to matrices in B_1. This is also so for $\hat{\varphi}$ and $\psi \ge \varphi$ in B_1 so that $R(\hat{\varphi}) \le R\psi$. ∎

4. A microstructure that is not a laminate

The description of this gradient parametrized measure that is not a laminate requires the following fact which is a refinement of the definition of a laminate. It refers to the support of the matrices that participate in the set of pairs verifying (H_l) conditions.

Lemma 9.9 *Let ν be a laminate supported in a compact set K with first moment $Y \in co(K)$. There exists a sequence of sets of pairs $\{(t_i^k, Y_i^k)\}_{1\le i\le k}$ satisfying the (H_k) condition such that $Y_i^k \in co(K)$ for all k and i, and*

$$\sum_i t_i^k \delta_{Y_i^k} \overset{*}{\rightharpoonup} \nu. \tag{9–8}$$

Proof. By Definition 9.3, there exist sets of pairs $\{(s_j^k, Z_j^k)\}_{1\le j\le k}$ verifying the (H_k) condition and

$$\sum_j s_j^k \delta_{Z_j^k} \overset{*}{\rightharpoonup} \nu.$$

In particular all matrices Z_j^k are contained in some fixed ball. If we write down what the (H_k) condition means, we can find a decomposition of the set of indices $\{1, 2, \dots, k\}$ in two proper subsets T_i^k, $i = 1, 2$ such that if

$$t_i^k = \sum_{T_i^k} s_j^k, \quad Y_i^k = \sum_{T_i^k} \frac{s_j^k}{t_i^k} Z_j^k, \quad i = 1, 2,$$

then $\operatorname{rank}(Y_1^k - Y_2^k) \le 1$. After taking some subsequence in k, let us set $Y_i^k \to Y_i$, $t_i^k \to t_i \in [0,1]$, $i = 1, 2$. We claim that if $Y_i \notin co(K)$, then $t_i = 0$. Suppose not, and find an open, convex neighborhood of $co(K)$, U, such that $Y_i \notin U$. For k sufficiently large, $Y_i^k \notin U$ and $t_i^k \ge t_i/2 > 0$. This implies, due to the convexity of U, that the sum

$$\sum_{j\in T_i^k, Z_j^k \notin U} \frac{s_j^k}{t_i^k}$$

does not converge to 0. Since the denominators t_i^k are uniformly distant from 0, we conclude

$$\sum_{j\in T_i^k, Z_j^k \notin U} s_j^k$$

does not converge to 0, either. This is a contradiction with the support of ν contained in U. If $t_i = 0$ discard the pair (t_i, Y_i). In any case $Y = t_1 Y_1 + t_2 Y_2$.

We now proceed recursively. By the definition of the (H_k) condition the set of pairs $\{(s_j^k, Z_j^k)\}_{j\in T_1^k}$ verifies some (H_l) condition, and therefore we can apply the above procedure to this set of pairs, finding a further subsequence for the indices k and pairs (t_{1i}, Y_{1i}), $i = 1, 2$, such that $\operatorname{rank}(Y_{11} - Y_{12}) \le 1$, $Y_1 = t_{11} Y_{11} + t_{12} Y_{12}$ and $Y_{1i} \in co(K)$ provided $t_{1i} > 0$. Keep performing these decompositions for T_2^k and subsequent subsets. We clearly see that for an appropriate diagonal subsequence and renaming the coefficients t_i and the matrices Y_i as t_i^l and Y_i^l, we have

$$\lim_{l\to\infty}\lim_{k\to\infty}\left|\sum_i t_i^l\psi(Y_i^l)-\sum_j s_j^k\psi(Z_j^k)\right|=0,$$

$$\left\{\left(t_i^l,Y_i^l\right)\right\}_{1\le i\le l}\text{ satisfies }(H_l),\quad Y_i^l\in\operatorname{co}(K),$$

for any continuous function. This implies that the weak limit for

$$\sum_i t_i^l\delta_{Y_i^l}\text{ and }\sum_j s_j^k\delta_{Z_j^k}$$

is the same. This is the end of the proof. ∎

The counterexample is constructed by means of a periodic deformation. Let $\chi=(2\chi_0-1)$ where χ_0 is the characteristic function of $(0,1/2)$ in $(0,1)$ extended by periodicity. We define a deformation $u:\Omega=(0,1)^2\subset\mathbf{R}^2\to\mathbf{R}^3$ by putting

$$u_1(x)=\int_0^{x_1}\chi(s)\,ds,$$
$$u_2(x)=\int_0^{x_2}\chi(s)\,ds,$$
$$u_3(x)=\int_0^{x_1+x_2}\chi(s+\frac{1}{4})\,ds.$$

The gradient ∇u is the matrix

$$\nabla u(x)=\begin{pmatrix}\chi(x_1) & 0\\ 0 & \chi(x_2)\\ \chi(x_1+x_2+\frac14) & \chi(x_1+x_2+\frac14)\end{pmatrix}$$

Notice that the gradient always lies in the three-dimensional subspace, L, of matrices of the form

$$\begin{pmatrix}x & 0\\ 0 & y\\ z & z\end{pmatrix}\mapsto(x,y,z).\tag{9–9}$$

If we consider the sequence of Ω-periodic functions $u_j(x)=(1/j)u(jx)$, by the Riemann-Lebesgue lemma (Lemma 8.2) the gradient parametrized measure corresponding to the sequence of gradients $\{\nabla u_j\}=\{\nabla u(jx)\}$ is homogeneous and, after the identification suggested in (9–9), given by

$$\begin{aligned}\nu=&\frac{1}{16}\left[\delta_{(1,1,1)}+\delta_{(1,-1,-1)}+\delta_{(-1,1,-1)}+\delta_{(-1,-1,1)}\right]\\ &+\frac{3}{16}\left[\delta_{(1,1,-1)}+\delta_{(1,-1,1)}+\delta_{(-1,1,1)}+\delta_{(-1,-1,-1)}\right].\end{aligned}$$

The determination of the different weights for ν is a matter of careful counting. Let V denote the set of vertices of the cube $C=[-1,1]^3$, so that $\operatorname{supp}(\nu)=V$ and the first moment is the origin. We claim that ν is not a laminate. Indeed we have:

Proposition 9.10 *If μ is a laminate with $\operatorname{supp}(\mu) = V$ and first moment 0,*

$$\mu = \frac{1}{8}\left[\delta_{(1,1,1)} + \delta_{(1,-1,-1)} + \delta_{(-1,1,-1)} + \delta_{(-1,-1,1)}\right] + \frac{1}{8}\left[\delta_{(1,1,-1)} + \delta_{(1,-1,1)} + \delta_{(-1,1,1)} + \delta_{(-1,-1,-1)}\right].$$

The proof reduces to the observation that by Lemma 9.9, in trying to find laminates supported on the set of vertices V and having first moment 0, we can restrict ourselves to the cube C. Since **the only rank-one directions in C are the ones given by the axes**, the only possibility is the one claimed by the proposition.

The situation when the dimension for the target space $m = 2$ is drastically distinct. A natural extension of a periodic deformation that might aid in finding a counterexample would be $u : \Omega = (0,1)^2 \to \mathbf{R}^2$ defined by

$$u_1(x) = \int_0^{x_1} \chi(s)\, ds + \int_0^{x_1+x_2} \chi(s + \frac{1}{4})\, ds,$$

$$u_2(x) = \int_0^{x_2} \chi(s)\, ds + \int_0^{x_1+x_2} \chi(s + \frac{1}{4})\, ds.$$

The gradient of u lies now in the three-dimensional subspace of 2×2 symmetric matrices. The underlying gradient parametrized measure generated from u by homogenization can be represented through the identification

$$\begin{pmatrix} x+z & z \\ z & y+z \end{pmatrix} \mapsto (x, y, z),$$

and again by

$$\nu = \frac{1}{16}\left[\delta_{(1,1,1)} + \delta_{(1,-1,-1)} + \delta_{(-1,1,-1)} + \delta_{(-1,-1,1)}\right] + \frac{3}{16}\left[\delta_{(1,1,-1)} + \delta_{(1,-1,1)} + \delta_{(-1,1,1)} + \delta_{(-1,-1,-1)}\right].$$

What changes now is the set of rank-one directions contained in the corresponding three-dimensional subspace L. Indeed vectors $(x, y, z) \in L$ yield rank-one directions if and only if

$$\det \begin{pmatrix} x+z & z \\ z & y+z \end{pmatrix} = xy + xz + yz = 0.$$

We claim that ν is now a laminate. Consider the points in the cube $C = [-1,1]^3 \subset L$ whose coordinates are given below

$$P_0 = (0,0,0), \quad P_1 = (-\frac{1}{2}, 1, 1),$$

$$P_2 = (\frac{1}{10}, -\frac{1}{5}, -\frac{1}{5}), \quad P_3 = (1, -\frac{5}{7}, 1),$$

$$P_4 = (-\frac{1}{11}, -\frac{1}{11}, -\frac{5}{11}), \quad P_5 = (1, 1, -1),$$

$$P_6 = (-1, -1, 0).$$

It is easy to check that

$$P_2 - P_1, P_4 - P_3, P_6 - P_5$$

are all rank-one (we mean that the coordinates of those differences satisfy $xy + yz + xz = 0$).

We can build a laminate using these directions, supported on the set of vertices of the cube in the following way. Write first

$$P_0 = \lambda_1 P_2 + (1 - \lambda_1) P_1,$$

where $P_2 - P_1$ is a rank-one direction and $\lambda \in (0, 1)$ is chosen appropriately. Likewise

$$P_2 = \lambda_2 P_3 + (1 - \lambda_2) P_4,$$
$$P_1 = \lambda_3 (1, 1, 1) + (1 - \lambda_3)(-1, 1, 1),$$

where once again $P_4 - P_3$ and $(1, 1, 1) - (-1, 1, 1)$ are rank-one directions and $\lambda_2, \lambda_3 \in (0, 1)$. Hence

$$P_0 = \lambda_1 [\lambda_2 P_3 + (1 - \lambda_2) P_4] + (1 - \lambda_1) [\lambda_3 (1, 1, 1) + (1 - \lambda_3)(-1, 1, 1)] .$$

In the same way

$$P_3 = \lambda_4 (1, 1, 1) + (1 - \lambda_4)(1, -1, 1),$$
$$P_4 = \lambda_5 P_5 + (1 - \lambda_5) P_6,$$
$$P_6 = \frac{1}{2}(-1, -1, 1) + \frac{1}{2}(-1, -1, -1),$$

where $(1, 1, 1) - (1, -1, 1)$, $P_5 - P_6$, $(-1, -1, 1) - (-1, -1, -1)$ are rank-one directions. Use these decompositions to find a laminate supported in V. The laminate, ν_1, that comes out of this construction is

$$\begin{aligned} \nu_1 = & \frac{3}{16} \left(\delta_{(-1,-1,-1)} + \delta_{(-1,-1,1)}\right) \\ & + \frac{2}{16} \left(\delta_{(1,-1,1)} + \delta_{(-1,1,1)}\right) \\ & + \frac{1}{16} \left(\delta_{(1,1,1)} + 5\delta_{(1,1,-1)}\right) . \end{aligned}$$

By symmetry we can also construct very easily two more laminates, ν_2 and ν_3, with the origin as first moment and with volume fractions given by

$$\begin{aligned}\nu_2 = &\frac{3}{16}\left(\delta_{(-1,-1,-1)} + \delta_{(-1,1,-1)}\right)\\ &+ \frac{2}{16}\left(\delta_{(-1,1,1)} + \delta_{(1,1,-1)}\right)\\ &+ \frac{1}{16}\left(\delta_{(1,1,1)} + 5\delta_{(1,-1,1)}\right),\\ \nu_3 = &\frac{3}{16}\left(\delta_{(-1,-1,-1)} + \delta_{(1,-1,-1)}\right)\\ &+ \frac{2}{16}\left(\delta_{(1,1,-1)} + \delta_{(1,-1,1)}\right)\\ &+ \frac{1}{16}\left(\delta_{(1,1,1)} + 5\delta_{(-1,1,1)}\right).\end{aligned}$$

It is a matter of careful arithmetic to find

$$\nu = \frac{1}{3}\nu_1 + \frac{1}{3}\nu_2 + \frac{1}{3}\nu_2.$$

Because of the convexity of the set of laminates with the same first moment, ν is a laminate.

5. Rank-one convexity does not imply quasiconvexity

Even though the arguments in the preceding section are convincing they may be confusing to readers with not much experience with laminates. The notation might also be somewhat confusing. We intend in this section to give an explicit counterexample of a rank-one convex function that is not quasiconvex. The reader will immediately see the connection with the example in the last section. Indeed, we are going to produce a rank-one convex function Ψ such that

$$\Psi(0) > \langle \Psi, \nu\rangle$$

where

$$\begin{aligned}\nu = &\frac{1}{16}\left[\delta_{(1,1,1)} + \delta_{(1,-1,-1)} + \delta_{(-1,1,-1)} + \delta_{(-1,-1,1)}\right]\\ &+ \frac{3}{16}\left[\delta_{(1,1,-1)} + \delta_{(1,-1,1)} + \delta_{(-1,1,1)} + \delta_{(-1,-1,-1)}\right],\end{aligned}$$

in the context of the previous section for $m = 3$. Therefore Ψ cannot be quasiconvex and ν cannot be a laminate.

Rather that dealing with ν directly we will use the periodic deformation, $u : \Omega = (0,1)^2 \subset \mathbf{R}^2 \to \mathbf{R}^3$ that generates it

$$u(x) = \left(\int_0^{x_1} \chi(s)\, ds, \int_0^{x_2} \chi(s)\, ds, \int_0^{x_1+x_2} \chi(s + \frac{1}{4})\, ds\right),$$

where $\chi = (2\chi_0 - 1)$ and χ_0 is the characteristic function of $(0,1/2)$ in $(0,1)$ extended by periodicity. Recall that the gradient of u is given by

$$\nabla u(x) = \begin{pmatrix} \chi(x_1) & 0 \\ 0 & \chi(x_2) \\ \chi(x_1 + x_2 + \frac{1}{4}) & \chi(x_1 + x_2 + \frac{1}{4}) \end{pmatrix}$$

and its image lies in the three-dimensional subspace of 3×2 matrices in (9–9).

The first lemma gives a reformulation of quasiconvexity in terms of periodic deformations. It is very convenient in this form.

Lemma 9.11 *A continuous function $\varphi : \mathbf{M}^{m\times N} \to \mathbf{R}$ is quasiconvex if and only if*

$$\int_{(0,1)^N} \varphi(Y + \nabla u(x))\, dx \geq \varphi(Y), \tag{9–10}$$

for every matrix Y and every smooth, periodic deformation $u : \mathbf{R}^N \to \mathbf{R}^m$.

Proof. If (9–10) holds for every periodic deformation u then φ is trivially quasiconvex since any $U \in W_0^{1,\infty}(\Omega)$ for $\Omega = (0,1)^N$ can be extended by periodicity.

Assume now that φ is quasiconvex and let u be any periodic deformation. Consider the sequence $u_j(x) = Yx + (1/j)u(jx)$, $x \in \Omega$. $\nabla u_j = Y + \nabla u(jx)$ and by the Riemann-Lebesgue lemma, Lemma 8.2, the gradient parametrized measure associated to $\{\nabla u_j\}$ is homogeneous and given by

$$\langle \varphi, \nu \rangle = \int_\Omega \varphi(Y + \nabla u(x))\, dx,$$

(Notice that $|\Omega| = 1$.) Hence, if φ is quasiconvex, Jensen's inequality

$$\int_\Omega \varphi(Y + \nabla u(x))\, dx \geq \varphi\left(Y + \int_\Omega \nabla u(x)\, dx\right) = \varphi(Y)$$

should hold. We have used the fact that if u is periodic then the integral of its gradient over one period cell vanishes. ∎

The basic ingredient for our counterexample is the cubic polynomial ψ defined on the subspace L, through the identification in (9–9), by

$$\psi(x, y, z) = xyz.$$

Because the only rank-one directions contained in L are the directions given by the axes, ψ is certainly rank-one convex in this subspace (indeed rank-one affine). Moreover, it is easy to check

$$\int_\Omega \psi(\nabla u(x))\, dx = -\frac{1}{8} < 0 = \psi(0,0,0). \tag{9–11}$$

Our task is to extend ψ to all of the space of 3×2 matrices so that it remains rank-one convex and the inequality (9–11) is preserved.

Let $\mathbf{M}$ stand for the space of 3×2 matrices and π for the orthogonal projection onto L. For a matrix $X \in \mathbf{M}$, $|X|$ is the usual euclidean norm.

Lemma 9.12 *For $\epsilon > 0$ fixed, there exists $k = k(\epsilon) > 0$ such that the function*

$$\Psi(X) = \psi(\pi X) + \epsilon\left|X\right|^2 + \epsilon\left|X\right|^4 + k\left|X - \pi X\right|^2 \tag{9–12}$$

is rank-one convex on $\mathbf{M}$.

Proof. Since Ψ is smooth, we have to check that

$$\left.\frac{d^2}{dt^2}\Psi(Y + tF)\right|_{t=0} \geq 0$$

holds for every matrix Y and every rank-one convex matrix F. We first compute

$$\begin{aligned}\left.\frac{d^2}{dt^2}\Psi(Y + tF)\right|_{t=0} &= \left.\frac{d^2}{dt^2}\psi(\pi Y + t\pi F)\right|_{t=0} \\ &\quad + 2\epsilon\left|F\right|^2 + 4\epsilon\left|Y\right|^2\left|F\right|^2 + 8\epsilon(Y \cdot F)^2 \\ &\quad + 2k\left|F - \pi F\right|^2,\end{aligned} \tag{9–13}$$

where $F \cdot Y$ is the dot product for matrices. The function $\psi(\pi X)$ is a third degree, homogeneous polynomial so that we can find $c > 0$ with

$$\left.\frac{d^2}{dt^2}\psi(\pi Y + t\pi F)\right|_{t=0} \geq -c\left|Y\right|\left|F\right|^2,$$

for all matrices Y, F. From (9–13) we have

$$\left.\frac{d^2}{dt^2}\Psi(Y + tF)\right|_{t=0} \geq \left(-c\left|Y\right| + 4\epsilon\left|Y\right|^2\right)\left|F\right|^2,$$

for all matrices Y, F. In particular if $|Y| \geq c/(4\epsilon)$,

$$\left.\frac{d^2}{dt^2}\Psi(Y + tF)\right|_{t=0} \geq 0 \tag{9–14}$$

for all F. On the other hand we also see from (9–13) that

$$\left.\frac{d^2}{dt^2}\Psi(Y + tF)\right|_{t=0} \geq \left.\frac{d^2}{dt^2}\psi(\pi Y + t\pi F)\right|_{t=0} + 2\epsilon\left|F\right|^2 + 2k\left|F - \pi F\right|^2, \tag{9–15}$$

for all Y, F. Let $h(Y, F, k)$ denote the right-hand side in (9–15) for $\epsilon > 0$ fixed. h is a continuous function of all its arguments. If we let

$$\mathcal{K} = \left\{(Y, F) \in \mathbf{M} \times \mathbf{M} : |Y| \leq \frac{c}{4\epsilon}, \operatorname{rank}(F) = 1, |F| = 1\right\},$$

a compact set of pairs of matrices, we claim that k_0 can be found so that $h(Y, F, k_0) > \epsilon$ for all pairs $(Y, F) \in \mathcal{K}$. Otherwise we could find a sequence $(Y_k, F_k) \in \mathcal{K}$ such that $h(Y_k, F_k, k) \leq \epsilon$. If, by compactness of $\mathcal{K}$, $(Y_k, F_k) \to (Y, F) \in \mathcal{K}$ as $k \to \infty$, it follows by (9–15), that $F = \pi F \in L$ and

$$\frac{d^2}{dt^2}\psi(\pi Y + t\pi F)\bigg|_{t=0} \leq -\epsilon.$$

This contradicts the fact that ψ is rank-one convex in L. Hence there exists k_0 such that

$$\frac{d^2}{dt^2}\Psi(Y + tF)\bigg|_{t=0} > \epsilon$$

for each pair $(Y, F) \in \mathcal{K}$. This, together with (9–14), proves the lemma. ∎

The following theorem is a direct consequence of the previous results.

Theorem 9.13 *There exist $\epsilon > 0$ and $k > 0$ such that Ψ given in (9–12) is rank-one convex but not quasiconvex.*

Proof. For the periodic deformation u considered above, we can choose $\epsilon > 0$ sufficiently small so that by (9–11) and because ∇u is uniformly bounded,

$$\int_{(0,1)^2} \left(\psi(\nabla u) + \epsilon\,|\nabla u|^2 + \epsilon\,|\nabla u|^4\right)\,dx < 0.$$

For this value of ϵ and by Lemma 9.12 we can find k such that the corresponding Ψ in (9–12) is rank-one convex. However,

$$\int_{(0,1)^2} \Psi(\nabla u)\,dx = \int_{(0,1)^2} \left(\psi(\nabla u) + \epsilon\,|\nabla u|^2 + \epsilon\,|\nabla u|^4\right)\,dx < 0 = \Psi(0).$$

By Lemma 9.11 Ψ is not quasiconvex. ∎

This example can be extended to the situation where the target space for deformation is at least three-dimensional $u : \mathbf{R}^N \to \mathbf{R}^m$, $m \geq 3$. The question of whether rank-one convexity implies quasiconvexity for $m = 2$ and any N is still open.

Corollary 9.14 *For $m \geq 3$, the function $\overline{\Psi}(X) = \Psi(TX)$ where T is the first 3×2 submatrix of X is rank-one convex but not quasiconvex.*

The proof reduces to the observation that the linear transformation T takes rank-one lines into rank-one lines so that $\overline{\Psi}$ is rank-one convex. It is enough to consider the periodic deformation $U : \mathbf{R}^N \to \mathbf{R}^m$ defined by

$$U(x) = (u_1(x_1, x_2), u_2(x_1, x_2), u_3(x_1, x_2), 0, \ldots, 0),$$

where $x \in \mathbf{R}^N$ $(N \geq 2)$, and $u = (u_1, u_2, u_3)$ is the previous periodic deformation.

6. Bibliographical remarks

The main counterexample of this chapter (Theorem 9.13) is contained in [298]. We have taken the material in Section 5 from that paper. The point of view in terms of laminates has been developed in [262]. Several attempts to find a counterexample when deformations take values in $\mathbf{R}^2$ have been carried out. See [100], [254], [264]. Numerical experiments have been tried out as well: [162]. The question, however, remains open in this case despite these efforts.

A systematic treatment of laminates can be found in [256], [257]. The characterization in terms of rank-one convex functions is also contained in these works although these facts were known to specialists before. The material in some papers is important for this approach: [95], [96]. [307] includes a complete discussion on the example with four matrices of Section 2.

There is a whole set of explicit examples and counterexamples related to the different notions of convexity. [8], [10], [19], [101], [219], [287], [295], [296], [299], [307], [308], [318] include some of those.

Some other interesting works treating topics such as quasiconvexity, rank-one connections, and so on are [32], [64], [232].

Chapter 10
Analysis of Divergence-Free Parametrized Measures

1. Introduction

The question we address in this chapter is the characterization of parametrized measures coming from sequences of vector-valued functions $u_j : \Omega \subset \mathbf{R}^N \to \mathbf{R}^m$ uniformly bounded in $L^\infty(\Omega)$ for which we have additional information in the form

$$\{Au_j\} \text{ relatively compact in } H^{-1}(\Omega), \tag{10–1}$$

for A a differential operator of type

$$(Au)_i = \sum_{l,k} a_{ilk} \frac{\partial u^l}{\partial x_k}, \quad i = 1, \ldots, s,$$

with constant coefficients a_{ilk}. In the previous chapters, we have concentrated on the fundamental case when $A = $ curl, m replaced by $m \times N$,

$$\operatorname{curl} u = \frac{\partial u_l^i}{\partial x_k} - \frac{\partial u_k^i}{\partial x_l}, \quad i = 1, \ldots, m, \quad l, k = 1, \ldots, N,$$

and we asked for $Au = 0$ rather than (10–1). In general, let

$$\vartheta = \{(\lambda, \xi) \in \mathbf{R}^m \times \mathbf{R}^N : \sum_{l,k} a_{ilk} \lambda_l \xi_k = 0\},$$

and let the characteristic cone be defined by

$$\Lambda = \left\{\lambda \in \mathbf{R}^m : \text{ there is a } \xi \in \mathbf{R}^N - \{0\}, (\lambda, \xi) \in \vartheta\right\}.$$

The significance of Λ relative to weak lower semicontinuity is contained in the following fact, which yields a necessary condition on the energy density.

Theorem 10.1 *Consider the functional*

$$I(u) = \int_\Omega \varphi(u(x))\, dx,$$

*defined for some particular class of competing functions. If I is weak * lower semicontinuous in $L^\infty(\Omega)$ for sequences $\{u_j\}$ such that (10–1) holds, then φ should be convex along the directions of the characteristic cone of A, Λ, i.e.,*

$$\varphi(ta + (1-t)b) \leq t\varphi(a) + (a-t)\varphi(b), \quad t \in [0,1],$$

provided that $a - b \in \Lambda$.

Proof. The idea of the proof is already contained in the proof of Lemma 8.6. If $a - b \in \Lambda$, let ξ be nonvanishing such that the pair $(a-b, \xi) \in \vartheta$. Define the sequence

$$u_j(x) = \frac{1}{j}\chi_t(jx\cdot\xi)(a-b) + b, \quad x \in \Omega,$$

where once again χ_s denotes the characteristic function of the interval $(0,s)$ as a subinterval of $(0,1)$ extended periodically to all of $\mathbf{R}$. Then

$$\frac{\partial u_j^l}{\partial x_k} = \chi_t'(jx\cdot\xi)\xi_k(a-b)^l,$$

where χ_t' should be understood in the distributional sense. On the other hand it should now be clear that u_j takes on the values a and b in alternate layers of relative width t and $1-t$ respectively, and that $u_j \overset{*}{\rightharpoonup} u$ in $L^\infty(\Omega)$ where u is the constant $ta + (1-t)b$. Therefore for all j

$$\sum_{l,k} a_{ilk}\frac{\partial u_j^l}{\partial x_k} = \chi_t'(jx\cdot\xi)\sum_{l,k} a_{ilk}(a-b)^l\xi_k = 0,$$

and by hypothesis

$$\int_\Omega \varphi(u)\, dx \leq \liminf_{j\to\infty}\int_\Omega \varphi(u_j)\, dx.$$

Recalling the properties of u_j, this inequality becomes exactly the claimed convexity. ∎

This condition is far from being sufficient. Indeed, notice that the characteristic cone corresponding to $A =$ curl is precisely the rank-one cone and, as we know, convexity along the rank-one directions does not force weak lower semicontinuity in the gradient case.

Our basic assumption on the operator A in the present chapter is that Λ is all of $\mathbf{R}^m$, and we want to show that in this case there are no restrictions on the family of parametrized measures coming from sequences bounded in $L^\infty(\Omega)$ that satisfy also (10–1). This is not surprising because of the following corollary.

Corollary 10.2 *Let $\varphi : \mathbf{R}^n \to \mathbf{R}^*$ and the operator A be given such that Λ is all of $\mathbf{R}^n$. The weak lower semicontinuity property*

$$\liminf_{j\to\infty} \int_\Omega \varphi(u_j)\, dx \geq \int_\Omega \varphi(u)\, dx,$$

holds whenever $u_j \overset{*}{\rightharpoonup} u$ in $L^\infty(\Omega)$ and $\{Au_j\}$ is relatively compact in $H^{-1}(\Omega)$, if and only if φ is convex.

In other words, condition (10–1) does not give additional weak lower semicontinuous integrands other than the convex ones. For the proof, simply observe that by Theorem 10.1 and our basic assumption, φ should be convex along any direction. Conversely, if φ is convex, the weak lower semicontinuity holds in general even without (10–1).

The most important operator under this assumption is div:

$$u : \Omega \subset \mathbf{R}^N \to \mathbf{R}^{N\times m}, \quad u = (u_i)_1^m, \quad \operatorname{div}(u) = (\operatorname{div}(u_i))_1^m,$$

for $m < N$.

The principal aim of this chapter is to prove the following fact.

Theorem 10.3 *Let the operator A be given with the property that its characteristic cone is all of space, and let $\nu = \{\nu_x\}_{x\in\Omega}$ be a family of probability measures depending measurably on $x \in \Omega$ such that $\operatorname{supp}(\nu_x) \subset K$ for a fixed compact $K \subset \mathbf{R}^m$. There exists a sequence $u^j : \Omega \subset \mathbf{R}^N \to K$ such that $\{Au^j\}$ is compact in $H^{-1}(\Omega)$ and its parametrized measure is ν.*

The technique to prove this theorem follows along the same line of thought as in the previous chapters. For this reason the present chapter is technical in nature, and we will not comment any further on the ideas behind the proof of Theorem 10.3. We just point out that the arguments essentially rely on two facts. First of all, in order to cut off functions in $L^\infty(\Omega)$ appropriately, we will use consistently the following decomposition for $\varphi \in H^1(\Omega)$ and η a cut-off function in Ω,

$$\int_\Omega u^j \eta A^* \varphi\, dx = \int_\Omega u^j A^* \eta\, \varphi\, dx - \int_\Omega u^j A^*(\eta\varphi)\, dx,$$

where A^* is the adjoint operator. The first term can be controlled by the weak convergence of $\{u^j\}$, and the second one by the compactness of $\{Au^j\}$ in $H^{-1}(\Omega)$.

Secondly, the fact that Λ is all of $\mathbf{R}^m$ will play a crucial role in Theorem 10.6 below.

From now on, we deal with $A = \operatorname{div}$ and $m = 1$ in the above description, since the arguments depend on the two facts just mentioned, and proofs might be obscured were we to use A instead of div.

2. Technical preliminaries

The basic lemma (and in particular the main idea in its proof) that we are going to use repeatedly is the following.

Lemma 10.4 *Let $\Omega \subset \mathbf{R}^N$ be any domain, bounded or unbounded, and $\chi = \chi_D$ the characteristic function of a Lipschitz subdomain $D \subset \Omega$. Assume $u^k \overset{*}{\rightharpoonup} u$ in $L^\infty(\Omega)$ and $\operatorname{div} u^k \to \operatorname{div} u$ in $H^{-1}(D)$. Then*

$$\operatorname{div}(\chi u^k) \to \operatorname{div}(\chi u) \text{ in } H^{-1}(\Omega).$$

In particular, $\operatorname{div}(\chi u^k) \to 0$ in $H^{-1}(\Omega)$, if $u \cdot n = 0$ in $\partial D \cap \Omega$, where n is the unit normal to D.

Observe that Ω could even be all of $\mathbf{R}^N$.
Proof. For $\varphi \in H_0^1(\Omega)$, writing

$$\begin{aligned}\int_\Omega \chi u^k \nabla\varphi\, dx &= \int_\Omega \chi(u^k - u)\nabla\varphi\, dx + \int_\Omega \chi u \nabla\varphi\, dx\\ &= \int_\Omega \chi(u^k - u)\nabla\varphi\, dx + \int_{\partial D\cap\Omega} u\cdot n\, \varphi\, dx,\end{aligned}$$

we can assume $u = 0$.

Let η_δ be a smooth cut-off function for D such that

$$\begin{gathered}0 \le \eta_\delta \le 1,\\ \eta_\delta = 0 \quad \text{in } \Omega - D,\\ \eta_\delta = 1 \quad \text{if } \operatorname{dist}(x, \partial D) \ge \delta,\ x \in D\\ |\nabla\eta_\delta| \le \frac{C}{\delta}.\end{gathered}$$

For $\varphi \in H_0^1(\Omega)$, $\|\nabla\varphi\|_{L^2(\Omega)} \le 1$, write

$$\int_\Omega \chi u^k \nabla\varphi\, dx = \int_\Omega (\chi - \eta_\delta) u^k \nabla\varphi\, dx - \int_D u^k \nabla\eta_\delta\, \varphi\, dx + \int_D u^k \nabla(\eta_\delta\varphi)\, dx.$$

We now control each of these terms on the right-hand side.
i) Given $\epsilon > 0$, choose $\delta > 0$, so that

$$\|\chi - \eta_\delta\|_{L^2(\Omega)} \le \frac{\epsilon}{C}, \quad \text{where } \left\|u^k\right\|_{L^\infty(\Omega)} \le C, \quad \text{for all } k.$$

ii) Once $\delta > 0$ is fixed, since $u^k \nabla\eta_\delta \rightharpoonup 0$ in $L^2(\Omega)$, the linear functionals

$$\langle T_k, \varphi\rangle = \int_\Omega u^k \nabla\eta_\delta\, \varphi\, dx, \quad \varphi \in H_0^1(\Omega),$$

converge strongly to 0 in $H^{-1}(\Omega)$; therefore, there is a k_0, such that if $k \geq k_0$

$$|\langle T_k, \varphi \rangle| \leq \epsilon, \quad \text{unif. in } \varphi.$$

iii) There is a k_1, such that if $k \geq k_1$

$$\| \operatorname{div} u^k \|_{H^{-1}(D)} \leq \frac{\epsilon}{\|\nabla(\eta_\delta \varphi)\|_{L^2(D)}}.$$

Hence if $k \geq \max(k_0, k_1)$,

$$\| \operatorname{div} (\chi u^k) \|_{H^{-1}(\Omega)} \leq 3\epsilon. \qquad \blacksquare$$

Another elementary device we need in this context is the average formula which we now establish as a lemma. It yields the homogenized version of any parametrized measure with the additional property on the divergences.

Lemma 10.5 *Assume that $u^k \overset{*}{\rightharpoonup} u$ in $L^\infty(\Omega)$, where $u \in \mathbf{R}^N$ is constant, and $\operatorname{div} u^k \to 0$ in $H^{-1}(\Omega)$. Let $\nu = \{\nu_x\}_{x \in \Omega}$ be the parametrized measure associated to $\{u^k\}$ and define*

$$\langle \overline{\nu}, \varphi \rangle = \frac{1}{|\Omega|} \int_\Omega \int_{\mathbf{R}^N} \varphi(\lambda)\, d\nu_x(\lambda) dx.$$

There exists $\{\overline{u}_k\}$ such that $\overline{u}_k \overset{*}{\rightharpoonup} u$ in $L^\infty(\Omega)$, $\operatorname{div} \overline{u}_k \to 0$ in $H^{-1}(\Omega)$, and the associated parametrized measure is $\overline{\nu}$.

Proof. Given j, let η_j be a cut-off function for Ω:

$$0 \leq \eta_j \leq 1,$$
$$\eta_1 = 1, \text{ if } \operatorname{dist}(x, \partial\Omega) \geq \frac{1}{j},$$
$$\eta_j = 0 \text{ on } \partial\Omega,$$
$$|\nabla \eta_j| \leq Cj.$$

For $k \in \mathbf{N}$, $\{a + \epsilon\overline{\Omega} : a \in \Omega, \epsilon < k^{-1}\}$ is a covering of Ω. By the Vitali's covering lemma there exists a countable family $\{a_i, \epsilon_i\}$ and a null-measure set N, such that

$$\Omega = \bigcup_i (a_i + \epsilon_i \overline{\Omega}) \cup N.$$

Define

$$\overline{u}^{k,j}(x) = \begin{cases} u + \left(u^k\left(\frac{x - a_i}{\epsilon_i}\right) - u\right) \eta_j\left(\frac{x - a_i}{\epsilon_i}\right), & x \in a_i + \epsilon_i \overline{\Omega} \\ u, & \text{otherwise.} \end{cases}$$

For $\varphi \in H_0^1(\Omega)$, $\|\nabla\varphi\|_{L^2(\Omega)} \leq 1$,

$$\int_\Omega \overline{u}^{k,j}\ \nabla\varphi\, dx = \sum_i \epsilon_i^N \int_\Omega \left(u^k(y) - u\right) \eta_j(y) \nabla\varphi\left(a_i + \epsilon_i y\right)\, dy.$$

If

$$\Phi(y) = \sum_i \epsilon_i^N \left(\frac{1}{\epsilon_i}\varphi\left(a_i + \epsilon_i y\right) - \alpha_i\right),$$

where $\alpha_i = \int_\Omega \frac{1}{\epsilon_i}\varphi(a_i + \epsilon_i y)\, dy$, then

$$\nabla\Phi(y) = \sum_i \epsilon_i^N \nabla\varphi(a_i + \epsilon_i y),$$

$$\|\nabla\Phi\|_{L^2(\Omega)} \leq \|\nabla\varphi\|_{L^2(\Omega)},$$

and

$$\begin{aligned}\int_\Omega \overline{u}^{k,j}\nabla\varphi\, dx &= \int_\Omega \left(u^k(y) - u\right) \eta_j(y) \nabla\Phi(y)\, dy \\ &= \int_\Omega \left(u^k - u\right) \nabla(\eta_j\Phi)\, dy - \int_\Omega \left(u^k - u\right) \nabla\eta_j\ \Phi\, dy.\end{aligned} \tag{10–2}$$

Similar to the proof of the previous lemma, given j, we can choose k_j, so as to make the two terms on the right-hand side of (10–2) arbitrarily small. If $\overline{u}^j = \overline{u}^{k_j,j}$,

$$\operatorname{div}\overline{u}^j \to 0 \text{ in } H^{-1}(\Omega).$$

Now for $\xi \in \mathcal{C}(\Omega)$ and $\varphi \in \mathcal{C}(\mathbf{R}^N)$,

$$\begin{aligned}\int_\Omega \varphi\left(\overline{u}^j(x)\right) \xi(x)\, dx &= \sum_i \epsilon_i^N \int_\Omega \varphi\left(\left(u^{k_j}(y) - u\right)\eta_j(y) + u\right) \xi\left(a_i + \epsilon_i y\right)\, dy \\ &= \sum_i \epsilon_i^N \xi\left(\overline{a}_i\right) \int_\Omega \varphi\left(\left(u^{k_j} - u\right)\eta_j + u\right)\, dy,\end{aligned}$$

for suitable $\overline{a}_i \in a_i + \epsilon_i\, \Omega$, so that

$$\int_\Omega \varphi\left(\overline{u}_j(x)\right) \xi(x)\, dx \to \int_\Omega \xi\, dx \int_\Omega \int_{\mathbf{R}^N} \varphi(\lambda)\, d\nu_x(\lambda) dx. \tag{10–3}$$

Indeed, by the uniform continuity of φ,

$$\begin{aligned}&\left|\int_\Omega \varphi\left(\left(u^{k_j} - u\right)\eta_j + u\right)\, dy - \int_\Omega \varphi\left(u^{k_j}\right)\, dy\right| \\ &\qquad \leq M \int_\Omega \left|\left(\eta_j - 1\right)\left(u^{k_j} - u\right)\right|\, dy \\ &\qquad \leq MC \int_{\Omega - \{\eta_j = 1\}} dy.\end{aligned}$$

and this term tends to 0 as $j \to \infty$. By density, (10–3) implies that $\left\{\varphi\left(\overline{u}^j\right)\right\}$ tends weakly * to the constant $\langle\overline{\nu}, \varphi\rangle$, and hence the parametrized measure associated to $\left\{\overline{u}^j\right\}$ is $\overline{\nu}$, constant throughout Ω. ■

3. Convexification

For a continuous function $\Phi : \mathbf{R}^N \to \mathbf{R}$ and a constant vector $u \in \mathbf{R}^N$, we define the div-convexification by

$$\Phi^{div}(u) = \inf\left\{ \lim_{j\to\infty} \frac{1}{|Q|} \int_Q \Phi\left(u^j\right) dx : u^j \overset{*}{\rightharpoonup} u \text{ in } L^\infty(Q), \right.$$
$$\left. \operatorname{div} u^j \to 0 \text{ in } H^{-1}(Q) \right\},$$

where Q is a bounded open domain with $|\partial Q| = 0$.

Clearly enough,
$$\Phi^{**} \le \Phi^{div} \le \Phi.$$

Theorem 10.6 *Φ^{div} is convex.*

Proof. Let $u = \lambda a + (1-\lambda) b$, $\lambda \in (0,1)$. Let u_a^k, u_b^k be such that $u_a^k \overset{*}{\rightharpoonup} a$, $u_b^k \overset{*}{\rightharpoonup} b$ in $L^\infty(\Omega)$, $\operatorname{div} u_a^k$, $\operatorname{div} u_b^k \to 0$ in $H^{-1}(\Omega)$. Apply Lemma 10.5 so that we may assume that the associated parametrized measure is homogeneous. Let ζ be the characteristic function of $(0,\lambda)$ in (0,1) extended by periodicity and $m \in \mathbf{R}^N$, $|m| = 1$, $m\cdot(a-b) = 0$ (this is the only place where we are using the condition that the characteristic cone for the operator div is all of space). Fix j, and let $D_j = \{x \in \Omega : \zeta(jm\cdot x) = 1\}$. Apply Lemma 8.1 to D_j and $u_a^k - u_b^k$. Observe $(a-b)\cdot n = 0$ in $\partial\Omega_j \cap \Omega$. Then

$$\operatorname{div}\left(\chi_{D_j}\left(u_a^k - u_b^k\right)\right) \to 0 \text{ in } H^{-1}(\Omega).$$

Choose k_j such that

$$\left\| \operatorname{div}\left(\chi_{D_j}\left(u_a^{k_j} - u_b^{k_j}\right)\right)\right\|_{H^{-1}(\Omega)} \le \frac{1}{j}.$$

Finally, set

$$u^j(x) = u_b^{k_j}(x) + \zeta(jm\cdot x)\left(u_a^{k_j}(x) - u_b^{k_j}(x)\right), \quad x \in \Omega.$$

We claim the following:

i) $\operatorname{div} u^j \to 0$ in $H^{-1}(\Omega)$;

ii) assume $u^j \overset{*}{\rightharpoonup} \overline{u}$ in $L^\infty(\Omega)$, and let $B \subset \Omega$ be a ball. Take $\{k_j\}$ even larger so that

$$\left| \int_{B\cap D_j} u_a^{k_j}(x)\, dx - |D_j \cap B|\, a \right| \le \frac{1}{2j},$$
$$\left| \int_{(\Omega - D_j)\cap B} u_b^{k_j}(x)\, dx - |(\Omega - D_j)\cap B|\, b \right| \le \frac{1}{2j}.$$

It is easy to check that

$$\int_B u^j \, dx \to |B| \, u.$$

Since we can identify a function in $L^\infty(\Omega)$ through its averages on a countable family of balls, choose $\{k_j\}$ even larger, if necessary, so that $u^j \overset{*}{\rightharpoonup} u$ in $L^\infty(\Omega)$ and

$$\begin{aligned}
\Phi^{div}(u) &\leq \lim_j \frac{1}{|\Omega|} \int_\Omega \Phi(u^j) \, dx \\
&= \frac{1}{|\Omega|} \lim_j \left(\int_{\Omega \cap D_j} \Phi\left(u_a^{k_j}\right) dx + \int_{\Omega \setminus D_j} \Phi\left(u_b^{k_j}\right) dx \right) \\
&= \frac{1}{|\Omega|} \lim_j \left(\frac{|\Omega \cap D_j|}{|\Omega|} \int_\Omega \Phi(u_a^{k_j}) \, dx + \frac{|\Omega \setminus D_j|}{|\Omega|} \int_\Omega \Phi(u_b^{k_j}) \, dx \right) \\
&= \lambda \lim_j \frac{1}{|\Omega|} \int_\Omega \Phi(u_a^{k_j}) \, dx + (1-\lambda) \lim_j \frac{1}{|\Omega|} \int_\Omega \Phi(u_b^{k_j}) \, dx.
\end{aligned}$$

Notice we have used the fact that the parametrized measures associated to both $\left\{u_a^{k_j}\right\}$ and $\left\{u_b^{k_j}\right\}$ are homogeneous. Since $\left\{u_a^k\right\}$ and $\left\{u_b^k\right\}$ were arbitrary,

$$\Phi^{div}(u) \leq \lambda \Phi^{div}(a) + (1-\lambda)\Phi^{div}(b).$$

■

Corollary 10.7 $\Phi^{div} = \Phi^{**}$.

For the proof, merely observe that

$$\Phi^{**} = \sup\{\varphi : \varphi \text{ is convex and } \varphi \leq \Phi\} \geq \Phi^{div} \geq \Phi^{**}.$$

A couple of important remarks:

1. Φ^{div} does not depend on Q, as a consequence of Corollary 10.7.
2. A careful look at the proof of Theorem 10.6 shows that, $(\hat{\Phi})^{div} = (\hat{\Phi})^{**}$, where

$$\hat{\Phi} = \begin{cases} \Phi, & \text{in } B \\ +\infty, & \text{in } \mathbf{R}^N - B \end{cases}$$

for any ball B. This is left to the interested reader.

4. Characterization of divergence-free parametrized measures

Let K be a compact set in $\mathbf{R}^N$ and fix a ball $B = B(0, M) \subset \mathbf{R}^N$, $M > 0$, containing K. $\text{co}(K)$ represents the usual convexification of K.

Lemma 10.8 *For $u \in co(K)$, the set of probability measures*

$$\mathcal{A}_u = \left\{ \nu \in C(B)' : \operatorname{supp}\nu \subset K, \int_K \lambda \, d\nu = u, \ \nu \text{ is associated to } \left\{u^k\right\}, \right.$$
$$\left. u^k : \Omega \to K, \ \operatorname{div} u^k \to 0 \text{ in } H^{-1}(\Omega) \right\}$$

is convex and weak- closed.*

Proof. The weak * closedness part is easy using appropiate subsequences since $\mathcal{C}(B)$ and $L^1(\Omega)$ are separable.

Let $\nu_i \in A_u$, $\lambda \in (0,1)$ and $u_i^k \overset{*}{\rightharpoonup} u$ in $L^\infty(\Omega)$, $\operatorname{div} u_i^k \to 0$ in $H^{-1}(\Omega), i = 1,2$.

Take $D \subset \Omega$, smooth with $|D| = \lambda |\Omega|$. Apply Lemma 10.4, and have

$$\operatorname{div} \left(\chi_D \left(u_1^k - u_2^k\right)\right) \to 0 \text{ in } H^{-1}(\Omega).$$

For $u^k = u_2^k + \chi_D \left(u_1^k - m_2^k\right)$, apply Lemma 10.5 and let the corresponding parametrized measure for $\left\{u^k\right\}$ be $\lambda\nu_1 + (1-\lambda)\nu_2$. ■

Theorem 10.9

$$\mathcal{A}_u = \left\{ \nu \in \mathcal{C}(B)' : \nu \text{ is a probability measure }, \operatorname{supp}\nu \subset K, \right.$$
$$\left. \int_K \lambda \, d\nu(\lambda) = u \right\}.$$

Proof. Assume that $\varphi \in \mathcal{C}(B)$ is such that

$$\langle \varphi, \nu \rangle \geq 0, \quad \forall \nu \in \mathcal{A}_u.$$

Then

$$\lim_{j\to\infty} \int_\Omega \hat{\varphi}(u^j) \, dx \geq 0,$$

whenever

$$u^j \overset{*}{\rightharpoonup} u \text{ in } L^\infty(\Omega),$$
$$\operatorname{div} u^j \to 0 \text{ in } H^{-1}(\Omega),$$

so that

$$(\hat{\varphi})^{div}(u) \geq 0.$$

Thus, by the final remark in the previous section,

$$(\hat{\varphi})^{**}(u) = (\hat{\varphi})^{div}(u) \geq 0,$$

and for any μ, probability measure with first moment u and support in K

$$\langle\varphi,\mu\rangle \geq \int_{\mathbf{R}^N} (\hat{\varphi})^{**}\, d\mu \geq (\hat{\varphi})^{**}(u) \geq 0.$$

By Hahn-Banach,

$$\overline{\operatorname{co}(\mathcal{A}_u)} = \mathcal{A}_u = \left\{\mu : \operatorname{supp}\mu \subset K, \int_K \lambda\, d\mu(\lambda) = u\right\}. \qquad \blacksquare$$

We now prove a nonhomogeneous version of Theorem 10.9 which consists in "patching" sequences appropriately. This is the natural argument when going from the homogeneous version of some fact to the non-homogeneous as we have already done before.

Theorem 10.10 *Let $\nu = \{\nu_x\}_{x\in\Omega}$ be a family of probability measures with $supp(\nu_x) \subset K$ and let*

$$u(x) = \int_K \lambda\, d\nu_x(\lambda) \in L^\infty(\Omega).$$

There exists a sequence $\{u^j\}$, $u^j : \Omega \to K$, $\operatorname{div} u^j \to \operatorname{div} u$ in $H^{-1}(\Omega)$ and such that

$$\int_\Omega \varphi(x, u^j(x))\, dx \to \int_\Omega \int_{\mathbf{R}^N} \varphi(x,\lambda)\, d\nu_x(\lambda)\, dx,$$

for every Carathéodory function φ.

Proof. Let N be the complement of the set of the points $a \in \Omega$ where

$$\left(\int_{a+\epsilon\Omega} |u(x) - u(a)|^2\, dx\right)^{1/2} \to 0, \quad \epsilon \to 0.$$

Let $r_k(a) > 0$, such that if $\epsilon < r_k(a)$ then

$$\int_{a+\epsilon\Omega} |u(x) - u(a)|^2\, dx \leq \frac{1}{k^2},$$

and consider a countable set of products $\zeta(x)\psi(\lambda)$ whose linear combination are dense in $L^1(\Omega; \mathcal{C}(\overline{B_M}))$ where as before $\overline{B_M} = \{|\lambda| \leq M\} \subset \mathbf{R}^N$ and $K \subset \overline{B_M}$.

Apply Lemma 7.9 and find $\{a_{ki}\}$, $\{\epsilon_{ki}\}$ with the appropriate properties. Choose a sequence η_k of smooth cut-off functions such that

$$\eta_k = 1 \text{ in } \Omega_k = \left\{x \in \Omega : \operatorname{dist}(x, \partial\Omega) \geq \frac{1}{k}\right\},$$
$$\eta_k = 0 \text{ on } \partial\Omega,$$
$$|\nabla\eta_k| \leq 2k.$$

Let $\{u_x^j\}$ generate ν_x, with $u_x^j : \Omega \to K$, $\operatorname{div} u_x^j \to 0$ in $H^{-1}(\Omega)$ and $u_x^j \overset{*}{\rightharpoonup} 0$ in $L^\infty(\Omega)$. Define the functions $u^{k,j}$ by

$$\begin{cases} \eta_k\left(\frac{x-a_{ki}}{\epsilon_{ki}}\right) u^j_{a_{ki}}\left(\frac{x-a_{ki}}{\epsilon_{ki}}\right) + \left(1-\eta_k\left(\frac{x-a_{ki}}{\epsilon_{ki}}\right)\right) u(a_{ki}), & x \in a_{ki}+\epsilon_{ki}\Omega \\ u(x), & \text{otherwise.} \end{cases}$$

For $\varphi \in H_0^1(\Omega)$, $\|\nabla\varphi\|_{L^2(\Omega)} \le 1$,

$$\begin{aligned} \int_\Omega \left(u(x) - u^{k,j}(x)\right) \nabla\varphi(x)\, dx =& \sum_i \int_{a_{ki}+\epsilon_{ki}\Omega} \left(u(x) - u(a_{ki})\right) \nabla\varphi(x)\, dx \\ &+ \sum_i \int_\Omega \left(u(a_{ki}) - u^j_{a_{ki}}(y)\right) \nabla\left(\eta_k \varphi_{ki}\right) dy \\ &- \sum_i \int_\Omega \left(u(a_{ki}) - u^j_{a_{ki}}(y)\right) \nabla\eta_k\ \varphi_{ki}\, dy \\ =& I_k + II_k + III_k, \end{aligned}$$

where

$$\varphi_{ki}(x) = \epsilon_{ki}^N \frac{1}{\epsilon_{ki}} \varphi(a_{ki} + \epsilon_{ki} x) - \alpha_{ki}.$$

and α_{ki} are constants to be chosen in a moment. Observe that

$$\left(\int_\Omega |\nabla\varphi_{ki}(x)|^2\, dx\right)^{1/2} = \left(\int_{a_{ki}+\epsilon_{ki}\Omega} |\nabla\varphi(y)|^2\, dy\right)^{1/2} \le 1,$$

and α_{ki} is chosen so that, by Poincaré's inequality, the H^1-norm of all such φ_{ki} are uniformly bounded in $H^1(\Omega)$.

We now proceed to estimate the three terms I_k, II_k and III_k.

$$\begin{aligned} |I_k| &\le \sum_i \|\nabla\varphi\|_{L^2(a_{ki}+\epsilon_{ki}\Omega)} \|u - u(a_{ki})\|_{L^2(a_{ki}+\epsilon_{ki}\Omega)} \\ &\le \frac{1}{k} \sum_i \|\nabla\varphi\|_{L^2(a_{ki}+\epsilon_{ki}\Omega)} \\ &\le \frac{1}{k}. \end{aligned}$$

Once k, i are fixed, choose j large enough so that

$$\left\| \operatorname{div} u^j_{a_{ki}} \right\|_{H^{-1}(\Omega)} \le \frac{1}{k 2^i} \frac{1}{\|\nabla\left(\eta_k \varphi_{ki}\right)\|_{L^2(\Omega)}},$$

whence

$$|II_k| \le \frac{1}{k}.$$

Since

$$\left(u(a_{ki} - u^j_{a_{ki}}(x)\right) \nabla \eta_k(x) \overset{*}{\rightharpoonup} 0 \quad \text{in } L^\infty(\Omega), \quad j \to \infty,$$

choosing j even larger (depending on k, i), we may bring about

$$|III_k| \leq \frac{1}{k}.$$

Therefore, for an appropriate choice j_k and letting $\overline{u}^k = u^{j_k,k}$ we have

$$\operatorname{div} \overline{u}^k \to \operatorname{div} u \quad \text{in } H^{-1}(\Omega).$$

The integral representation is now standard, taking j_k even larger if necessary in order to have it for the chosen countable set.

Observe that the sequence $\{u^{k,j}\}$ as defined above may not take values in the set K. By definition, it does except for small sets. More precisely, if

$$A^k = \bigcup_i (a_{ki} + \epsilon_{ki} \{\eta_k < 1\}),$$

then

$$\begin{aligned} |A^k| &\leq \sum_i \epsilon_{ki}^N |\{\eta_k < 1\}| \\ &= |\Omega| \, |\{\eta_k < 1\}| \to 0 \end{aligned}$$

as $k \to \infty$ (notice that $\sum_i \epsilon_{ki}^N = |\Omega|$) and $u^{k,j}$ takes values in K outside A^k. If we define $\overline{u}^k$ to be some $u_0 \in K$, constant in A^k, the new sequence does not lose any of its interesting properties: the underlying parametrized measure does not change (Lemma 6.3) and the divergences of the new sequence is still a compact set in $H^{-1}(\Omega)$. ■

The proof of Theorem 10.10 could be reduced to the case $u \equiv 0$. This reduction, however, would require some technicalities concerning the support of the parametrized measure. We have preferred to give a direct proof.

Corollary 10.11 *Given any family of probability measures $\nu = \{\nu_x\}_{x\in\Omega}$, $\Omega \subset \mathbf{R}^N$, whose support is contained in a fixed compact K, there exists a sequence in $L^\infty(\Omega)$, $\{u^j\}$, such that*

i) $u^j : \Omega \to K$, for all j;

ii) $\{\operatorname{div}(\chi_\Omega u^j)\}$ is compact in $H^{-1}(\mathbf{R}^N)$;

iii) the parametrized measure associated to $\{u^j\}$ is ν.

Proof. By the theorem just proved, there exists a sequence, $\{u^j\}$, with properties i) and iii), and such that for

$$u(x) = \int_{\mathbf{R}^N} \lambda \, d\nu_x(\lambda),$$

we have

$$\operatorname{div} u^j \to \operatorname{div} u \quad \text{in } H^{-1}(\Omega).$$

By Lemma 10.4,

$$\operatorname{div}(\chi_\Omega u^j) \to \operatorname{div}(\chi_\Omega u) \quad \text{in } H^{-1}(\mathbf{R}^N). \qquad \blacksquare$$

Notice that Theorem 10.3 is an immediate consequence of this corollary.

5. Bibliographical remarks

We have already mentioned in Chapter 1 some basic sources for the theory of compensated compactness. The approach presented here based on parametrized measures can be found in the same format in [259]. Some of the technical tools used in this chapter were motivated by many of the works related to micromagnetism mentioned in Chapter 5.

Bibliography

[1] Abeyaratne, R., Chu, C. and James, R. 1994 Kinetics and hysteresis in martensitic single crystals, Proc. Symp. Mech. Phase Trans. Shape Mem. Alloys, ASME.

[2] Acerbi, E. and Buttazzo, G. 1984 Semicontinuous envelopes of polyconvex integrals, Proc. R. Soc. Edinb., 96A, 51–54.

[3] Acerbi, E., Buttazzo, G. and Fusco, N. 1983 Semicontinuity and relaxation for integrals depending on vector valued functions, J. Math. Pures Appl., 62, 371–387.

[4] Acerbi, E. and Fusco, N. 1984 Semicontinuity problems in the calculus of variations, Arch. Rat. Mech. Anal., 86, 125–145.

[5] Acerbi, E. and Fusco, N. 1987 A regularity theorem for minimizers of quasiconvex integrals, Arch. Rat. Mech. Anal., 99, 261–281.

[6] Adams, R. A. 1975 *Sobolev Spaces*, Academic Press, New York.

[7] Akhiezer, N. I. 1962 *The Calculus of Variations*, Blaisdell, New York.

[8] Albert, A. A. 1983 A quadratic form problem in the calculus of variations, Bull. AMS, 44, 250–253.

[9] Alibert, J. J. and Bouchitté, G. 1995 Non-uniform integrability and generalized Young measures, preprint.

[10] Alibert, J. J. and Dacorogna, B. 1992 An example of a quasiconvex function not polyconvex in dimension two, Arch. Rat. Mech. Anal., 117, 55–166.

[11] Allaire, G. and Kohn, R. V. 1996 Optimal lower bounds on the elastic energy of a composite made from two non well-ordered isotropic materials, Quart. Appl. Math., to appear.

[12] Ambrosio, L. 1987 New lower semicontinuity results for integral functionals, Rend. Accad. Naz. Sci. XL Mem. Mat. Sci. Fis. Natur., 105, 1–42.

[13] Ambrosio, L. 1989 Variational Problems in SBV, Acta Appl. Math., 17, 1–40.

[14] Ambrosio, L. 1990 Existence theory for a new class of variational problems, Arch. Rat. Mech. Anal., 111, n 4, 291–322.

[15] Ambrosio, L. and Buttazzo, G. 1988 Weak lower semicontinuous envelope of functionals defined on a space of measures, Ann. Mat. Pura Appl., 150, 311–340.

[16] Ambrosio, L., De Giorgi, E., 1988 Un Nuovo tipo di Funzionale del Calcolo delle Varizioni, Atti. Acc. Naz. dei Lincei, Rend. Cl. Sc. Fis. Mat. Natur. LXXXII, 199–210.

[17] Antman, S. 1983 The influence of elasticity on analysis: modern developments, Bull. AMS, 9, 267–291.

[18] Attouch, H. 1984 *Variational Convergence of Functionals and Operators*, Pitman, London.

[19] Aubert, G. 1987 A counterexample of a rank-one convex function which is not polyconvex in the case n 2, Proc. R. Soc. Edinb., 106A, 237–240.

[20] Aubert, G. and Tahraoui, R. 1979 Théorèmes d'existence pour des problèmes du calcul des variations, J. Diff. Eq., 33, 1–15.

[21] Aubert, G. and Tahraoui, R. 1980 Sur la minimisation d'une fonctionelle non convexe, non différentialble en dimension 1, Boll. Un. Mat. Ital., 17, 244–258.

[22] Aubert, G. and Tahraoui, R. 1983 Sur quelques résultats d'existence en optimisation non convexe, CRAS Paris, 297, 287–289.

[23] Aubert, G. and Tahraoui, R. 1984 Théorèmes d'existence en optimisation non convexe, Appl. Anal., 18, 75–100.

[24] Balder, E. J. 1984 A general approach to lower semicontinuity and lower closure in optimal control theory, SIAM J. Control and Opt., 22, 570–598.

[25] Balder, E. J. 1984 Existence results without convexity conditions for general problems of optimal control with singular components, J. Math. Anal. Appl., 101, 527–539.

[26] Balder, E. J. 1994 New existence results for optimal controls in the absence of convexity: the importance of extremality, SIAM J. Control Optim., 32, 890–916.

[27] Balder, E. J. 1995 Lectures on Young Measures, Cahiers de Ceremade, 9512.

[28] Balian, R. 1991 *From Microphysics to Macrophysics, I*, Springer-Verlag, Berlin.

[29] Ball, J. M. 1977 Convexity conditions and existence theorems in nonlinear elasticity, Arch. Rat. Mech. Anal., 63, 337–403.

[30] Ball, J. M. 1980 Strict convexity, strong ellipticity and regularity in the calculus of variations, Math. Proc. Camb. Phil. Soc., 87, 501–513.

[31] Ball, J. M. 1989 A version of the fundamental theorem for Young measures, *PDE's and Continuum Models of Phase Transitions*, Lecture Notes in Physics, 344,(Rascle, M. et al., eds.) Springer, 207–215.

[32] Ball, J. M. 1990 Sets of gradients with no rank-one connections, J. Math. Pures Appl., 69, 241–259.

[33] Ball, J. M., Currie, J. C. and Olver, P. J. 1981 Null Lagrangians, weak continuity, and variational problems of arbitrary order, J. Functional Anal., 41, 135–174.

[34] Ball, J. M. and James, R. D. 1987 Fine phase mixtures as minimizers of energy, Arch. Rat. Mech. Anal., 100, 13–52.

[35] Ball, J. M. and James, R. 1992 Proposed experimental tests of a theory of fine microstructure and the two well problem, Phil. Trans. R. Soc. London A, 338, 389–450.

[36] Ball, J. M., Knops, R. J. and Marsden, J. E. 1978 Two examples in nonlinear elasticity, in *Proceedings Conference on Nonlinear Analysis, Besançon*, 41–49, Lecture Notes in Mathematics, 466, Springer Verlag, Berlin.

[37] Ball, J. M. and Marsden, J. E. 1984 Quasiconvexity at the boundary, positivity of the second variation and elastic stability, Arch. Rat. Mech. Anal., 86, 251–277.

[38] Ball, J. M. and Murat, F. 1984 $W^{1,p}$-quasiconvexity and variational problems for multiple integrals, J. Fnal. Anal., 58, 225–253.

[39] Ball, J. M. and Murat, F. 1989 Remarks on Chacon's biting lemma, Proc. AMS, 107, 655–663.

[40] Ball, J. M. and Zhang, K. 1990 Lower semicontinuity of multiple integrals and the biting lemma, Proc. R. Soc. Edinb., 114A, 367–379.

[41] Barbu, V. 1989 *Mathematical Methods in Optimization of Differential Systems*, Kluwer Academic Publishers.

[42] Battacharya, K. 1991 Wedge-like microstructure in martensite, Acta Metal., 39, 2431–2444.

[43] Battacharya, K. 1992 Self accomodation in martensite, Arch. Rat. Mech. Anal., 120, 201–244.

[44] Battacharya, K., Firoozye, N., James, R. D. and Kohn, R. V. 1994 Restrictions on microstructure, Proc. R. Soc. Edinb., 124A, 843–878.

[45] Bhattacharya, K., James, R. and Swart, P. 1995 A nonlinear dynamic model for twin relaxation with applications to Au 47.5 AT. % Cd and other shape-memory alloys, preprint.

[46] Berliocchi, H. and Lasry, J. M. 1973 Intégrandes normales et mesures paramétrées en calcul des variations, Bull. Soc. Math. France, 101, 129–184.

[47] Billingsley, P. 1986 *Probability and Measure*, Wiley.

[48] Biskamp, D. 1993 *Nonlinear Magnetohydrodynamics*, Cambridge, England.

[49] Bonnetier, E. and Conca, C. 1993 Relaxation totale d'un problème d'optimisation de plaques, CRAS Paris, 317, 931–936.

[50] Bonnetier, E. and Conca, C. 1994 Approximation of Young measures by functions and application to a problem of optimal design for plates with variable thickness, Proc. R. Soc. Edinb., A, 124, 843–878.

[51] Bonnetier, E. and Vogelius, M. 1987 Relaxation of a compliance functional for a plate optimization problem, in *Applications of Multiple Scaling in Mechanics*, (P.G. Ciarlet et al., eds.), Masson, 31–53.

[52] Bouchitté, G. 1987 Représentation intégrale de fonctionnelles convexes sur un espace de mesures, Ann. Univ. Ferrara, 33, 113–156.

[53] Bouchitté, G. and Buttazzo, G. 1990 New lower semicontinuity results for nonconvex functionals defined on measures, Nonlinear Anal., 15, 679–692.

[54] Bouchitté, G. and Valadier, M. 1988 Integral representations of nonconvex functionals defined on measures, J. Funct. Anal., 80, 398–420.

[55] Brandon, D. and Rogers, R. 1992 The coercivity paradox and nonlocal ferromagnetism, Cont. Mech. Therm., 4, 1–21.

[56] Brezis, H. 1983 *Analyse Fonctionnelle*, Masson, Paris.

[57] Brighi, B. and Chipot, M. 1992 Approximated convex envelope of a function, preprint.

[58] Brooks, J. K. and Chacon, R. V. 1980 Continuity and compactness of measures, Adv. in Math., 37, 16–26.

[59] Brown, W. F. 1963 *Micromagnetics*, Wiley-Interscience, New York.

[60] Brown, W. F. 1966 *Magnetoelastic Interactions*, vol. 9 of Springer Tracts in Natural Philosophy (C. Truesdell, ed.), Springer Verlag, New York.

[61] Buttazzo, G. 1989 *Semicontinuity, Relaxation and Integral Representations in the Calculus of Variations*, Pitman Res. Notes Math. Ser., 207, Longman, Harlow.

[62] Buttazzo, G. and Dal Maso, G. 1985 Integral representation and relaxation of local functionals, Nonlinear Anal., 9, 515–532.

[63] Carstensen, C. and Plechac, P. 1995 Numerical solution of the scalar double-well problem allowing microstructure, preprint.

[64] Casadio Tarabusi, E., 1993 An algebraic characterization of quasiconvex functions, Ric. di Mat., vol. XLII, 11–24.

[65] Cellina, A. 1993 On minima of a functional of the gradient: necessary conditions, Nonlinear Anal., 20, 337–341.

[66] Cellina, A. 1993 On minima of a functional of the gradient: sufficient conditions, Nonlinear Anal., 20, 343–347.

[67] Cellina, A. and Colombo, G. 1990 On a classical problem of the calculus of variations, Ann. Inst. Henri Poincaré, Anal. Non Linéaire, 7, 97–106.

[68] Cellina, A. and Zagatti, S. 1996 An existence result for a minimum problem in the vectorial case of the calculus of variations, SIAM J. Control Opt., to appear.

[69] Cesari, L. 1974 An existence theorem without convexity conditions, SIAM J. Control, 12, 319–331.

[70] Cesari, L. 1983 *Optimization, Theory and Applications*, Springer-Verlag.

[71] Cheng, K. T. and Olhoff, N. 1981 An investigation concerning optimal design of solid elastic plates, Int. J. of Sol. Struc., 17, 305–323.

[72] Cheng, K. T. and Olhoff, N. 1982 Regularized formulation for optimal design of axisymmetric plates, Int. J. of Sol. Struc., 18, 153–169.

[73] Chipot, M. 1990 Hyperelasticity for crystals, Europ. J. of Appl. Math., 1, 113–129.

[74] Chipot, M. 1991 Numerical analysis of oscillations in nonconvex problems, Numer. Math., 59, 747–767.

[75] Chipot, M. and Collins, C. 1992 Numerical approximation in variational problems with potential wells, SIAM J. Numer. Anal., 29, 1002–1019.

[76] Chipot, M., Collins, C. and Kinderlehrer, D. 1995 Numerical analysis of oscillations in multiple well problems, Numer. Math., 70, 259–282.

[77] Chipot, M. and Kinderlehrer, D. 1988 Equilibrium configurations of crystals, Arch. Rat. Mech. Anal., 103, 237–277.

[78] Ciarlet, P. G. 1987 *Mathematical Elasticity*, vol I of Three-dimensional Elasticity, North-Holland, Amsterdam.

[79] Coifman, P. L., Lions, P. L., Meyer, Y. and Semmes, S. 1989 Compacité par compensation et espaces de Hardy, CRAS Paris, 309, 945–949.

[80] Coifman, P. L., Lions, P. L., Meyer, Y. and Semmes, S. 1993 Compensated Compactness and Hardy Spaces, J. Math. Pures Appl., 72, 247–286.

[81] Collins, C. 1993 Computation of twinning, in *Microstructure and Phase Transitions*, (Kinderlehrer, D. et al., eds.) IMA vol. 54, 39–50.

[82] Collins, C. and Luskin, M. 1989 The computation of the austenitic-martensitic phase transition, in *PDE's and Continuum Models of Phase Transitions*, Lecture Notes in Physics, 344, (Rascle, M. et al., eds.) Springer, 34–50.

[83] Collins, C. and Luskin, M. 1989 Computational results for phase transitions in shape memory materials, in *Smart Materials, Structures, and Mathematical Issues*, (C. Rogers, ed.), Technomic Publishing, 198–215.

[84] Collins, C. and Luskin, M. 1990 Numerical modeling of the microstructure of crystals with symmetry-related variants, in *Proc. ARO US-Japan Workshop on Smart/Intelligent Materials and Systems*, Technomic Publishing.

[85] Collins, C. and Luskin, M. 1991 Optimal order error estimates for the finite element approximation of the solution of a nonconvex variational problem, Math. Comp., 57, 621–637.

[86] Collins, C., Luskin, M. and Riordan, J. 1991 Computational images of crystalline microstructure, in *Computing Optimal Geometries* (J. Taylor, ed.) AMS Special Lectures in Mathematics and AMS Videotape Library, AMS, Providence, 16–18.

[87] Collins, C., Luskin, M. and Riordan, J. 1993 Computational results for a two-dimensional model of crystalline microstructure, in *Microstructure and Phase Transitions* (Kinderlehrer, D. et al., eds.) IMA vol. 54, 51–56.

[88] Collins, C., Kinderlehrer, D., and Luskin, M. 1991 Numerical approximation of the solution of a variational problem with a double well potential, SIAM J. Numer. Anal., 28, 321–333.

[89] Courant, R. 1962 *Calculus of Variations*, Courant Institute publications, New York.

[90] Courant, R. and Hilbert, D. 1966 *Methods of Mathematical Physics*, Wiley, New York.

[91] Dacorogna, B. 1981 A relaxation theorem and its applications to the equilibrium of gases, Arch. Rat. Mech. Anal., 77, 359–386.

[92] Dacorogna, B. 1982 Quasiconvexity and relaxation of nonconvex variational problems, J. Funct. Anal., 46, 102–118.

[93] Dacorogna, B. 1982 *Weak continuity and weak lower semicontinuity for non-linear functionals*, Springer Lecture Notes, vol. 922.

[94] Dacorogna, B. 1985 Relaxation for some dynamical problems, Proc. R. Soc. Edinb., 100A, 39–52.

[95] Dacorogna, B. 1985 Remarques sur les notions de polyconvexité, quasiconvexité et convexité de rang 1, J. Math. Pures Appl., 64, 403–438.

[96] Dacorogna, B. 1987 A characterization of polyconvex, quasiconvex and rank-one convex envelopes, Rend. Circolo Mat. Palermo, 15, 37–58.

[97] Dacorogna, B. 1987 Convexity of certain integrals of the calculus of variations, Proc. R. Soc. Edinb., 107A, 15–26.

[98] Dacorogna, B. 1989 *Direct Methods in the Calculus of Variations*, Springer.

[99] Dacorogna, B. and Botteron, B. 1990 Existence and nonexistence results for noncoercive variational problems and applications in ecology, J. Diff. Eq., 85, n 2, 214–235.

[100] Dacorogna, B., Douchet, J., Gangbo, W. and Rappaz, J. 1990 Some examples of rank-one convex functions in dimension two, Proc. R. Soc. Edinb., 114A, 135–150.

[101] Dacorogna, B. and Marcellini, P. 1988 A counterexample in the vectorial calculus of variations, in *Material Instabilities in Continuum Mechanics*, (J. M. Ball, ed.), 77–83, Oxford: Clarendon Press.

[102] Dacorogna, B. and Marcellini, P. 1995 Existence of minimizers for non-quasiconvex integrals, Arch. Rat. Mech. Anal., 131, 359–399.

[103] Dal Maso, G, 1993 *Introduction to* Γ*-convergence*, Boston MA, Birkhäuser.

[104] De Giorgi, E. 1985 Some continuity and relaxation problems, in *E. De Giorgi Colloquium, Proc.,* (P. Kree, ed.) Pitman Res. Notes, vol. 25, London, 1–11.

[105] De Giorgi, E., Ambrosio, L. and Buttazzo, G. 1987 Integral representation and relaxation for functionals defined on measures, Atti. Accad. Naz. Lincei Rend. Cl. Sci. Fis. Mat. Natur., 81, 7–13.

[106] De Giorgi, E., Buttazzo, G. and Dal Maso, G. 1983 On the lower semicontinuity of certain integral functionals, Alti Accad. Naz. Lincei Rend., 74, 274–282.

[107] Demoulini, S. 1996 Young measure solutions for a nonlinear parabolic equation of forward-backward type, SIAM Math. Anal., 27, n 2, 376–403.

[108] De Simone, A. 1993 Energy minimizers for large ferromagnetic bodies, Arch. Rat. Mech. Anal., 125, 99–143.

[109] De Simone, A. and Friesecke, G. 1996 On the problem of two linearized wells, Calc. Var., 4, 293–304.

[110] DiPerna, R. J. 1983 Convergence of approximate solutions to conservation laws, Arch. Rat. Mech. Anal., 82, 27–70.

[111] DiPerna, R. J. 1983 Convergence of the viscosity method for isentropic gas dynamics, Comm. Math. Phys., 91, 1–30.

[112] DiPerna, R. J. 1985 Compensated compactness and general systems of conservation laws, Trans. Amer. Math. Soc. 292, 383–420.

[113] DiPerna, R. J. and Majda, A. J. 1987 Oscillations and concentrations in weak solutions of the incompressible fluid equations, Comm. Math. Phys., 108, 667–689.

[114] DiPerna, R. J. and Majda, A. J. 1987 Concentrations in regularizations for 2-D incompressible flow, Comm. Pure Appl. Math., 40, 301–345.

[115] Dunford, N. and Schwartz, J. T. 1957 *Linear Operators*, Wiley, New York.

[116] Edwards, R. E. 1965 *Functional Analysis*, Holt, Rinehart and Winston.

[117] Ekeland, I. 1979 Nonconvex minimization problems, Bull. AMS, 1, 443–475.

[118] Ekeland, I. and Temam, R. 1976 *Convex Analysis and Varational Problems*, North-Holland, Amsterdam.

[119] Ellis, R. S. 1985 *Entropy, Large Deviations and Statistical Mechanics*, Springer-Verlag, New York.

[120] Ericksen, J. L., 1979 On the symmetry of deformable crystals, Arch. Rat. Mech. Anal., 72, 1–13.

[121] Ericksen, J. L. 1980 Some phase transitions in crystals, Arch. Rat. Mech. Anal., 73, 99–124.

[122] Ericksen, J. L. 1981 Changes in symmetry in elastic crystals, in *IUTAM Symp. Finite Elasticity*, (D. E. Carlson et al., eds.), 167–177, Nijhoff.

[123] Ericksen, J. L. 1981 Some simpler cases of the Gibbs phenomenon for thermoelastic solids, J. of Thermal Stresses, 4, 13–30.

[124] Ericksen, J. L. 1982 Crystals, lattices and sublattices, Rend. Sem. Mat. Padova, 68, 1–9.

[125] Ericksen, J. L. 1983 Ill posed problems in thermoelasticity theory, in *Systems of Nonlinear Partial Differential Equations*, (J. Ball, ed.), 71–95, Reidel.

[126] Ericksen, J. L. 1984 The Cauchy-Born hypotheses for crystals, in *Phase Transformations and Material Instabilities in Solids*, (M. Gurtin, ed.), 61–78, Academic Press.

[127] Ericksen, J. L. 1986 Constitutive theory for some constrained elastic crystals, Int. J. Solids Structures, 22, 951–964.

[128] Ericksen, J. L. 1986 Stable equilibrium configurations of elastic crystals, Arch. Rat. Mech. Anal., 94, 1–14.

[129] Ericksen, J. L. 1987 Twinning of crystals, in *Metastability and Incompletely Posed Problems*, IMA Vol. Math. Appl., vol. 3, (S. Antman et al., eds.), 77–96.

[130] Ericksen, J. L. 1988 Some constrained elastic crystals, in *Material Instabilities in Continuum Mechanics*, (J. Ball, ed.), 119–136, Oxford.

[131] Ericksen, J. L. 1989 Weak martensitic transformations in Bravais lattices, Arch. Rat. Mech. Anal., 107, 23–36.

[132] Evans, L. C., 1990 *Weak Convergence Methods for Nonlinear Partial Differential Equations*, CBMS 74, American Mathematical Society.

[133] Evans, L. C. and Gariepy, R. F. 1987 Blow-up, compactness, and partial regularity in the calculus of variations, Indiana Univ. Math. J., 36, 361–371.

[134] Evans, L. C. and Gariepy, R. F. 1987 Some remarks on quasiconvexity and strong convergence, Proc. R. Soc. Edinb., 106A, 53–61.

[135] Evans, L. C. and Gariepy, R. F. 1992 *Lecture Notes on Measure Theory and Fine Properties of Functions.* CRC Press.

[136] Federer, H. 1969 *Geometric Measure Theory*, Springer-Verlag, Berlin.

[137] Firoozye, N. B. 1991 Optimal use of the translation method and relaxation of variational problems, Comm. Pure Appl. Math., 44, 643–678.

[138] Firoozye, N. B. and Kohn, R. V. 1993 Geometric parameters and the relaxation of multiwell energies, in *Microstructure and Phase Transition*, (D. Kinderlehrer et al., eds.), IMA vol. 54, 85–110.

[139] Fonseca, I. 1985 Variational methods for elastic crystals, Arch. Rat. Mech. Anal., 97, 189–220.

[140] Fonseca, I. 1988 The lower quasiconvex envelope of the stored energy function for an elastic crystal, J. Math. Pures Appl., 67, 175–195.

[141] Fonseca, I. 1992 Lower semicontinuity of surface energies, Proc. R. Soc. Edinb., 120A, 99–115.

[142] Fonseca, I., Kinderlehrer, D., Pedregal, P. 1994 Energy functionals depending on elastic strain and chemical composition, Calc. Var., 2, 283–313.

[143] Fonseca, I., Müller, S. 1992 Quasiconvex integrands and lower semicontinuity in L^1, SIAM J. Math. Anal., 23, 1081–1098.

[144] Fonseca, I., Müller, S. 1993 Relaxation of quasiconvex functionals in $BV(\Omega\ \mathbf{R}^p)$ for integrands $f(x, u, \nabla u)$, Arch. Rat. Mech. Anal., 123, 1–49.

[145] Fonseca, I., Müller, S. and Pedregal, P. 1995 Analysis of concentration and oscillation effects generated by gradients, SIAM J. Math. Anal., submitted.

[146] Francfort, G. A. and Milton, G. W. 1994 Sets of conductivity and elasticity tensors stable under lamination, Comm. Pure Appl. Math., 47, 257–279.

[147] French, D. A. 1990 On the convergence of finite element approximations of a relaxed variational problem, SIAM J. Numer. Anal., 27, 419–436.

[148] Friesecke, G. 1994 A necessary and sufficient condition for nonattainment and formation of microstructure almost everywhere in scalar variational problems, Proc. R. Soc. Edinb. Section A, 124, 437–472.

[149] Friesecke, G. and Dolzmann, G. 1995 Implicit time discretization and global existence for a quasi-linear evolution equation with nonconvex energy, SIAM Math. Anal. (to appear).

[150] Friesecke, G. and McLeod J. B. 1996 Dynamics as a mechanism preventing the formation of finer and finer microstrucuture, Arch. Rat. Mech. Anal., 133, n 3, 199–247.

[151] Fusco, N. 1980 Quasiconvessità e semicontinuità per integrali multipli de ordine superiore, Ricerche Mat., 29, 307–323.

[152] Fusco, N. 1983 Remarks on the relaxation of integrals of the calculus of variations, in *Systems of Nonlinear PDE*, (J. M. Ball, ed.), Reidel, 401–408.

[153] Gamkrelidze, R. V. 1978 *Principles of Optimal Control Theory*, Plenum Press, New York.

[154] Gerard, P. 1988 *Compacité par compensation et regularité 2-microlocale*, Séminaire Eq. aux Dér. Part., Ecole Polytechnique, Palaiseau, exp VI.

[155] Gerard, P. 1989 Microlocal defect measures, Comm. Part. Diff. Eq., 16 (11), 1761–1794.

[156] Giachetti, D. and Schianchi, R. 1993 Minima of some non-convex, non-coercive problems, Ann. Mat. Pura. Appl., 165, 109–120.

[157] Giaquinta, M. 1983 *Multiple Integrals in the Calculus of Variations and Nonlinear Elliptic Systems*, Princeton Univ. Press, Princeton.

[158] Giaquinta, M. and Hildebrandt, S. 1996 *Calculus of Variations, I and II*, Springer.

[159] Giaquinta, M. and Modica, G. 1986 Parial regularity of minimizers of quasiconvex integrals, Anal. Non Linéaire, 3, 185–208.

[160] Gilbarg, D. and Trudinger, N. S. 1977 *Elliptic Partial Differential Equations of Second Order*, Springer, Berlin.

[161] Giusti, E. 1984 *Minimal Surfaces and Functions of Bounded Variation*, Birkhäuser, Boston.

[162] Gremaud, P. 1994 Numerical analysis of a nonconvex variational problem related to solid-solid phase transitions, SIAM J. Numer. Anal., 31, 111–127.

[163] Gurtin, M. E. 1981 *An Introduction to Continuum Mechanics*, Academic Press, New York.

[164] Halmos, P. R. 1950 *Measure Theory*, Van Nostrand, Princeton.

[165] Holm, D., Marsden, J., Ratiu, T. and Weinstein, A. 1985 Nonlinear stability of fluid and plasma equilibria, Phys. Rep., 123, 1–116.

[166] Ioffe, A. D. 1977 On lower semicontinuity of integral functionals I, II, SIAM J. Control Optim., 15, 521–538 and 991–1000.

[167] Iwaniec, T. and Lutoborski, A. 1993 Integral estimates for null Lagrangians, Arch. Rat. Mech. Anal., 125, n 1, 25–79.

[168] Iwaniec, T. and Sbordone, C. 1992 On the integrability of the Jacobian under minimal hypothesis, Arch. Rat. Mech. Anal., 119, 129–144.

[169] James, R. D. 1988 Microstructure and weak convergence, in *Proc. Symp. Material Instabilities in Cont. Mech.*, (J. M. Ball, ed.), 175–196.

[170] James, R. D. and Kinderlehrer, D. 1989 Theory of diffusionless phase transitions, in *PDE's and Continuum Models of Phase Transitions, Lecture Notes in Physics*, 344, (M. Rascle et al., eds.), 51–84.

[171] James, R. D. and Kinderlehrer, D. 1990 Frustration in ferromagnetic materials, Cont. Mech. Therm., 2, 215–239.

[172] James, R. D. and Kinderlehrer, D. 1993 A theory of magnetostriction with application to TbDyFe2, Phil. Mag., B 68, 237–274.

[173] James, R. D. and Kinderlehrer, D. 1990 Frustration and microstructure: An example in magnetostriction, in *Proc. First Europ. Conf. Elliptic Parab. Prob.*, Pont à Mousson.

[174] James, R. D. and Kinderlehrer, D. 1993 Mathematical Approaches to the Study of Smart Materials, Carnegie Mellon Univ. preprint.

[175] James, R., Kinderlehrer, D. and Ma, L. 1996 Modeling magnetostrictive microstructure under loading, in *Proc. TMS Symposium*, submitted.

[176] James, R. D. and Müller, S. 1994 Internal variables and fine-scale oscillations in micromagnetics, Cont. Mech. Therm., 6, n 4, 291–336.

[177] Jordan, R. K., 1994 *Statistical Equilibria and Coherent Structures in Two-dimensional Magnetohydrodynamic Turbulence*, PhD Thesis, University of Massachusetts, Amherst.

[178] Jordan, R. K., 1995 A statistical equilibrium model of coherent structures in magnetohydrodynomics, Nonlinearity, 8, 585–613.

[179] Jordan, R. K. and Kinderlehrer, D. 1996 An extended variational principle, in *Part. Diff. Eq. and Applications*, (Marcellini, P. et al., eds.). Lect. Notes Pure and Appl. Math. 177, Marcel Dekker, 187–200.

[180] Jordan, R. and Turkington, B. 1996 Ideal magnetofluid turbulence in two dimensions, submitted.

[181] Kinderlehrer, D. 1988 Remarks about the equilibrium configurations of crystals, in *Proc. Symp. Material Instabilities in Cont. Mech.*, (J. M. Ball ed.), 217–242, Oxford.

[182] Kinderlehrer, D. 1995 Stability of simulated magnetic systems, in *Proc. Equadiff-95*, (Magalhaes, L. et al., eds), submitted.

[183] Kinderlehrer, D. and Ma, L. 1995 The hysteretic event in the computation of magnetization, CMU preprint.

[184] Kinderlehrer, D. and Ma, L. 1996 Simulation of magnetoelastic systems, in *Proc. SPIE Math. and Control in Smart Struct.*, vol 2715, (V. Varadan, ed).

[185] Kinderlehrer, D., Nicolaides, R. A. and Wang, H. 1993 Spurios oscillations in computing microstructures, Carnegie Mellon Univ. preprint.

[186] Kinderlehrer, D. and Pedregal, P. 1991 Characterizations of Young measures generated by gradients, Arch. Rat. Mech. Anal., 115, 329–365.

[187] Kinderlehrer, D. and Pedregal, P. 1991 Charactérisation des mesures de Young associées à un gradient, CRAS Paris, 313, 765–770.

[188] Kinderlehrer, D. and Pedregal, P. 1994 Remarks about gradient Young measures generated by sequences in Sobolev spaces, in *Nonlinear PDE and Their Applications*, Collège de France Seminar, Vol. XI, Pitman Res. Notes, Longman, 131–150.

[189] Kinderlehrer, D. and Pedregal, P. 1992 Remarks about the analysis of gradient Young measures, in *PDE and Related Subjects*, (M. Miranda, ed.), Pitman-Longman Res. Notes, 269, 125–150.

[190] Kinderlehrer, D. and Pedregal, P. 1992 Weak convergence of integrands and the Young measure representation, SIAM J. Math. Anal., 23, 1–19.
[191] Kinderlehrer, D. and Pedregal, P. 1994 Gradient Young measures generated by sequences in Sobolev spaces, J. of Geom. Anal., 4, 59–90.
[192] Kinderlehrer, D. and Pedregal, P. 1996 Remarks about Young measures supported on two wells, to appear.
[193] Knops, R. J. and Stuart, C. A. 1984 Quasiconvexity and uniqueness of equilibrium solutions in nonlinear elasticity, Arch. Rat. Mech. Anal., 86, 233–249.
[194] Kohn, R. 1991 The relaxation of a double-well energy, Cont. Mech. Thermodyn., 3, 193–236.
[195] Kohn, R. V. and Müller, S. 1992 Branching of twins near an austenite/twinned martensite interface, Phil. Magazine A, 66, n 5, 697–715.
[196] Kohn, R. V. and Müller, S. 1994 Surface energy and microstructure in coherent phase transitions, Comm. Pure Appl. Math., 47, no 4, 405–435.
[197] Kohn, R. V. and Strang, G. 1987 Optimal design and relaxation of variational problems, I, II and III, Comm. Pure Appl. Math., 34, 113–137, 139–182 and 353–377.
[198] Kohn, R. V. and Vogelius, M. 1986 Thin plates with varying thickness, and their relation to structural optimization, in *Homogenization and Effective Moduli of Materials and Media*, IMA Volumes 1, (Ericksen, J. et al., eds.), Springer-Verlag, 126–149.
[199] Kohn, R. V. and Vogelius, M. 1987 Relaxation of a variational method for impedance computed tomography, Comm. Pure Appl. Math., 40, 745–777.
[200] Kristensen, J. 1994 Finite functionals and Young measures generated by gradients of Sobolev functions, Mathematical Institute, Technical University of Denmark, Mat-Report No. 1994–34.
[201] Li, Z. P. 1996 A theorem on lower semicontinuity of integral functionals, Proc. R. Soc. Edinb., 126A, 363–374.
[202] Li, B. and Luskin, M. 1996 Finite element analysis of microstructure for the cubic to tetragonal transformation, IMA preprint.
[203] Lions, P. L. 1984 The concentration-compactness principle in the calculus of variations: the locally compact case, part 1 and 2, Ann. Inst. H. Poincaré Anal. Non Linéaire, 1, 109–145, 223–283.
[204] Lions, P. L. 1985 The concentration-compactness principle in the calculus of variations. The limit case, part 1 and 2, Rev. Mat. Iberoam., vol. 1, n 1, 145–201; vol 1, n 2, 45–121.
[205] Luskin, M. 1995 Approximation of a laminated microstructure for a rotationally invariant, double well energy density, IMA preprint.
[206] Luskin, M. 1996 Approximation of a laminated microstructure for a rotationally invariant, double well energy density, Numerische Math., to appear.
[207] Luskin, M. 1996 On the computation of crystalline microstructure, Acta Numerica, 5, 191–257.

[208] Luskin, M. and Ma, L. 1992 Analysis of the finite element approximation of microstructure in micromagnetics, SIAM J. of Numer. Anal., 29, 320–331.

[209] Ma, L. 1993 Computation of magnetostrictive materials, Carnegie Mellon preprint.

[210] Ma, L. 1991 Analysis and computation for a variational problem in micromagnetics, Ph.D. Thesis, University of Minnesota, Minneapolis, MN.

[211] Ma, L. and Walkington, N. 1995 On algorithms for non-convex optimization in the calculus of variations, SIAM J. Numer. Anal., 32, n 3, 900–923.

[212] MacShane, E. J. 1940 Generalized curves, Duke Math. J., 6, 513–536.

[213] MacShane, E. J. 1940 Necessary conditions in the generalized curve problem of the calculus of variations, Duke Math. J., 7, 1–27.

[214] MacShane, E. J. 1947 *Integration*, Princeton Univ. Press.

[215] Malý, J. 1993 Weak lower semicontinuity of polyconvex integrals, Proc. R. Soc. Edinb., 123A, 681–691.

[216] Marcellini, P. 1979 Some problems of semicontinuity, in *Recent Methods in Nonlinear Analysis, Proceedings*, (E. De Giorgi et al., eds.), Pitagora, Bologna, 205–222.

[217] Marcellini, P. 1980 Alcune osservazioni sull'esistenza del minimo di integrali del calcolo delle variazioni senza ipotesi de convessità, Rend. Mat., 13, 271–281.

[218] Marcellini, P. 1983 A relation between existence of minima for nonconvex integrals and uniqueness for non strictly convex integrals of the calculus of variations, in *Mathematical Theories of Optimization, Proceedings*, (J. P. Cecconi et al., eds.), Springer Lecture Notes in Math., vol. 979, 216–231.

[219] Marcellini, P. 1984 Quasiconvex quadratic forms in two dimensions, Appl. Math. Optim., 11, 183–189.

[220] Marcellini, P. 1985 Approximation of quasiconvex functions and lower semcontinuity of multiple integrals, Manus. Math., 51, 1–28.

[221] Marcellini, P. 1986 On the definition and the lower semicontinuity of certain quasiconvex integrals, Ann. Inst. H. Poincaré: Anal. non Linéaire, 3, 391–409.

[222] Marcellini, P. 1989 Regularity of minimizers of integrals of the calculus of variations with nonstandard growth conditions, Arch. Rat. Mech. Anal., 105, n 3, 267–284.

[223] Marcellini, P. 1990 Nonconvex integrals of the calculus of variations, in *Methods of Nonconvex Analysis*, Springer Lecture Notes in Math., vol. 1446, 16–57.

[224] Marcellini, P. and Sbordone, C. 1977 Relaxation of nonconvex variational problems, Alti Acad. Naz. Lincei, 63, 341–344.

[225] Marcellini, P. and Sbordone, C. 1980 Semicontinuity problems in the calculus of variations, Nonlinear Anal., 4, 241–257.

[226] Marcellini, P. and Sbordone, C. 1983 On the existence of minima of multiple integrals of the calculus of variations, J. Math. Pure Appl., 62, 1–10.

[227] Mascolo, E. and Schianchi, R. 1983 Existence theorems for nonconvex problems, J. Math. Pures Appl., 62, 349–359.

[228] Mascolo, E. and Schianchi, R. 1985 Nonconvex problems of the calculus of variations, Nonlinear Anal., 9, 371–380.

[229] Mascolo, E. and Schianchi, R. 1987 Existence theorems in the calculus of variations, J. Diff. Eq., 67, 185–198.

[230] Matos, J. 1992 Young measures and the absence of fine microstructures in a class of phase transitions, Europ. J. Appl. Math., 3, 31–54.

[231] Matos, J. 1996 On gradient Young measures supported on a point and a well, preprint.

[232] Matousek, J. and Plechac, P. 1995 On functional separately convex hulls, preprint.

[233] Meyers, N. G. 1965 Quasiconvexity and lower semicontinuity of multiple variational integrals of any order, Trans. AMS, 119, 125–149.

[234] Miller, J. 1990 Statistical mechanics of Euler equations in two dimensions, Phys. Rev. Lett., 65, 2137–2140.

[235] Milton, G. W. 1990 On characterizing the set of possible effective tensors of composites: the variational method and the translation method, Comm. Pure Appl. Math., 42, 63–125.

[236] Milton, G. W. 1994 A link between sets of tensors stable under lamination and quasiconvexity, Comm. Pure Appl. Math., 47, 959–1003.

[237] Milton, G. W. and Nesi, V. 1991 Polycrystalline configurations that maximize electrical resistivity, J. Mech. Phys. Solids, 39, 525–542.

[238] Morrey, Ch. B., 1952 Quasiconvexity and the lower semicontinuity of multiple integrals, Pacific J. Math., 2, 25–53.

[239] Morrey, Ch. B. 1966 *Multiple Integrals in the Calculus of Variations*, Springer, Berlin.

[240] Müller, S. 1988 Weak continuity of determinants and nonlinear elasticity, CRAS Paris, 307, 501–506.

[241] Müller, S. 1989 A surprising higher integrability property of mappings with positive determinant, Bull. AMS, 21, 245–248.

[242] Müller, S. 1990 Higher integrability of determinants and weak convergence in L^1, J. Reine Angew. Math., 412, 20–34.

[243] Müller, S. 1990 Minimizing sequences for nonconvex functionals, phase transitions and singular perturbations, Springer Lecture Notes in Physics, 31–44.

[244] Müller, S. and Sverak, V. 1996 Attainment results for the two-well problem by convex integration, preprint.

[245] Muñoz, J. and Pedregal, P. 1996 On the relaxation of an optimal design problem for plates, Asympt. Anal., to appear.

[246] Murat, F. 1977 H-convergence, *Séminaire d'analyse fonctionelle et numérique de l'Université d'Alger.*

[247] Murat, F. 1978 Compacité par compensation, Ann. Scuola Norm. Sup. Pisa Sci. Fis. Mat. (IV), 5, 489–507.

[248] Murat, F. 1979 Compacité par compensation II, in *Recent Methods in Nonlinear Analysis, Proceedings*, (E. De Giorgi et al., eds.), Pitagora, Bologna, 245–256.

[249] Murat, F. 1987 A survey on compensated compactness, in *Contributions to the Modern Calculus of Variations*, (L. Cesari, ed.), Pitman, 145–183.

[250] Murat, F. and Tartar, L. 1985 Calcul des variations et homogénéisation, in *Les méthodes de l'homogénéisation: théorie et applications en physique*. Dir. des études et recherches de l'EDF, Eyrolles, Paris, 319–370.

[251] Nicolaides, R. A. and Walkington, N. 1992 Computation of microstructure utilizing Young measure representations, in *Proc. Recent Adv. Adaptive Sensory Materials and Their Appl.*, Blacksburg, VA.

[252] Nicolaides, R. A. and Walkington, N. 1992 Strong convergence of numerical solutions to degenerate variational problems, Carnegie Mellon Univ. preprint.

[253] Nicolaides, R. A., Walkington, N. and Wang, H. 1995 Numerical methods for a non-convex optimization problem modeling martensitic phase transitions, CNA preprint.

[254] Parry, G. P. 1995 On the planar rank-one convexity condition, Proc. R. Soc. Edinb., 125 A, 247–264.

[255] Pedregal, P. 1989 Weak continuity and weak lower semicontinuity for some compensation operators, Proc. R. Soc. Edinb., 113A, 267–279.

[256] Pedregal, P. 1993 The concept of a laminate, in *Microstructure and Phase Transitions*, IMA Volumes 54, (D. Kinderlehrer et al., eds.), 129–142.

[257] Pedregal, P. 1993 Laminates and Microstructure, Europ. J. Appl. Math., 4, 121–149.

[258] Pedregal, P. 1993 Remarks about separately convex functions, Appl. Math. Opt., 28, n 1, 1–10.

[259] Pedregal, P. 1994 Relaxation in ferromagnetism: the rigid case, J. of Nonlinear Science, 4, 105–125.

[260] Pedregal, P. 1994 Jensen's inequality in the calculus of vVariations, Diff. Int. Equat., 7, n 1, 57–72.

[261] Pedregal, P. 1995 Numerical computation of parametrized measures, Numer. Funct. Anal. Opt., 16, n 7–8, 1049–1066.

[262] Pedregal, P. 1996 Some remarks on quasiconvexity and rank-one convexity, Proc. R. Soc. Edinb., 126, n 5, 1055–1065.

[263] Pedregal, P. 1996 On the numerical analysis of nonconvex variational problems, Numerische Mathematik, 74, 325–336.

[264] Pedregal, P. and Sverak, V. 1996 The quasiconvexification of a cubic polynomial at the origin, in *Proc. Second World Congr. Nonlinear Anal.*, Athens, Greece (accepted).

[265] Rascle, D. 1986 Un résultat de compacité par compensation à coefficients variables. Application a l'elasticité non linéaire, CRAS Paris, 302, 311–314.

[266] Raymond, J. P. 1987 Conditions nécessaires et suffisantes d'existence de solution en calcul des variations, Anal. Non Linéaire, 4, 169–202.

[267] Raymond, J. P. 1992 Existence of minimizers for vector problems without quasiconvexity condition, Nonlinear Anal., 18, 815–828.

[268] Renardy, M. and Rogers, R. C. 1993 *An Introduction to Partial Differential Equations*, Springer.

[269] Reshetnyak, Y. 1967 General theorems on semicontinuity and on convergence with a functional, Sibir. Math., 8, 801–816.

[270] Reshetnyak, Y. 1967 On the stability of conformal mappings in multidimensional spaces, Sibir. Math., 8, 91–114.

[271] Reshetnyak, Y. 1968 Stability theorems for mappings with bounded excursion, Sibir. Math., 9, 667–684.

[272] Reshetnyak, Y. 1968 Weak convergence and completely additive vector functions on a set, Sibir. Math., 9, 1039–1045.

[273] Robert, R. 1991 A maximum entropy principle for two-dimensional perfect fluid dynamics, J. Stat. Phys., 65, 531–553.

[274] Rockafellar, R. T. 1972 *Convex Analysis*, Princeton University Press, Princeton.

[275] Rogers, R. C. 1988 Nonlocal variational problems in nonlinear electromagnetoelasticity, SIAM J. Math. Anal., 19, 1329–1347.

[276] Rogers, R. C. 1991 A nonlocal model for the exchange energy in ferromagnetic materials, J. Integral Eq. Appl., 3, 85–127.

[277] Roytburd, V. and Slemrod, M. 1988 An application of the method of compensated compactness to a problem in phase transitions, in *Material Instabilities in Continuum Mechanics*, (J. M. Ball, ed.), Oxford Univ. Press.

[278] Roubícek, T. 1995 Explicit characterization of L^p-Young measures, J. Math. Anal. Appl., to appear.

[279] Roubícek, T. 1995 Effective characterization of generalized Young measures generated by gradients, Boll. U. M. It., (7) 9-B, 755–779.

[280] Roubícek, T. 1995 A note about relaxation of vectorial variational problems, in *Calculus of Variations, Applications and Computations*, (C. Bandel et al., eds.) Pitman Res. Notes, 326, Longman, Harlow, Essex, 208–215.

[281] Roubícek, T. 1995 Numerical approximation of relaxed variational problems, J. on Convex Anal., to appear.

[282] Roubicek, T. 1996 *Relaxation in optimization theory and variational calculus*, W. De Gruyter, Berlin, to appear.

[283] Rudin, W. 1974 *Real and Complex Analysis*, McGraw Hill, New York.

[284] Rudin, W. 1973 *Functional Analysis*, McGraw Hill, New York.

[285] Saks, S. 1937 *Theory of the integral*, G. E. Stechert, New York.

[286] Schonbeck, M. E., 1982 Convergence of solutions to nonlinear dispersive equations, Comm. in Part. Diff. Eq., 7, 959–1000.

[287] Serre, D. 1983 Formes quadratiques et calcul des variations, J. Math. Pures Appl., 62, 177–196.

[288] Serre, D. 1987 La compacité par compensation pour les systèmes hyperboliques non linéaires à une dimension d'espace, J. Math. Pures Appl., 65, 423–468.

[289] Serrin, J. 1959 On a fundamental theorem of the calculus of variations, Acta Math., 102, 1–32.

[290] Slemrod, M. and Roytburd, V. 1987 Measure-valued solutions to a problem in dynamic phase transtitions, in *Contemporary Mathematics*, vol. 50, AMS.

[291] Stein, E. M. 1970 *Singular Integrals and Differentiability Properties of Functions*, Princeton University Press, Princeton.

[292] Sternberg, P. 1991 Vector valued local minimizers of nonconvex variational problems, Rocky Mountains J. Math., 21, 799–807.

[293] Strang, G. 1985 The polyconvexification of $F(\nabla u)$, Research Report CMA-RO 9-83 of the Australian National Univ., Canberra.

[294] Struwe, M. 1990 *Variational Methods*, Springer Verlag, Berlin.

[295] Sverak, V. 1990 Examples of rank-one convex functions, Proc. R. Soc. Edinb., 114A, 237–242.

[296] Sverak, V. 1991 Quasiconvex functions with subquadratic growth, Proc. R. Soc. Lond., A433, 723–725.

[297] Sverak, V. 1993 On the problem of two wells, in *Microstructure and Phase Transition* , (D. Kinderlehrer et al., eds.), 183–190, Springer Verlag.

[298] Sverak, V. 1992 Rank-one convexity does not imply quasiconvexity, Proc. R. Soc. Edinb., 120A, 185–189.

[299] Sverak, V. 1992 New examples of quasiconvex functions, Arch. Rat. Mech. Anal., 119, 293–300.

[300] Tahraoui, R. 1990 Sur une classe de fonctionnelles non convexes et applications, SIAM J. Math. Anal., 21, 37–52.

[301] Tartar, L. 1977 Cours Peccot, Collège de France.

[302] Tartar, L. 1979 Compensated compactness and applications to partial differential equations, in *Nonlinear analysis and mechanics: Heriot-Watt Symposium*, vol. IV, (R. Knops, ed.), Pitman Res. Notes Math., 39, 136–212.

[303] Tartar, L. 1983 The compensated compactness method applied to systems of conservation laws, in *Systems of Nonlinear Partial Differential Eq.*, (J. M. Ball, ed.), Riedel.

[304] Tartar, L. 1984 Ètude des oscillations dans les équations aux dérivées partielles nonlinéaires, Springer Lecture Notes in Physics, vol 195, 384–412.

[305] Tartar, L. 1990 H-measures, a new approach for studying homogenisation, oscillations and concentration effects in partial differential equations, Proc. R. Soc. Edinb., 115A, 193–230.

[306] Tartar, L. 1991 On mathematical tools for studying partial differential equations of continuum physics: H-measures and Young measures, in *De-*

velopments in Partial Differential Equations and Applications to Mathematical Physics, (G. Buttazzo et al., eds.), Plenum, New York.

[307] Tartar, L. 1993 Some remarks on separately convex functions, in *Microstructure and Phase Transitions*, (D. Kinderlehrer et al., eds.), IMA Volume 54, 191–204.

[308] Terpstra, F. J. 1938 Die Darstellung der biquadratischen Formen als Summen von Quadraten mit Anwendung auf die Variationsrechnung, Math. Ann., 116, 166–180.

[309] Tonelli, L. 1921 *Fondamenti di calcolo delle variazioni*, Zanichelli, Bologna.

[310] Troutman, J. L. 1983 *Variational Calculus with Elementary Convexity*, Springer, New York.

[311] Turkington, B. and Jordan, R. 1995 Turbulent relaxation of a magnetofluid: a statistical equilibrium model, to appear.

[312] Warga, J. 1962 Relaxed variational problems, J. Math. Anal. Appl., 4, 111–128.

[313] Warga, J. 1972 *Optimal Control of Differential and Functional Equations*, Academic Press, New York.

[314] Young, L. C. 1937 Generalized curves and the existence of an attained absolute minimum in the calculus of variations, Comptes Rendus de la Société des Sciences et des Lettres de Varsovie, classe III, 30, 212–234.

[315] Young, L. C. 1942 Generalized surfaces in the calculus of variations, I and II, Ann. Math., 43, 84–103 and 530–544.

[316] Young, L. C. 1969 *Lectures on Calculus of Variations and Optimal Control Theory*, W. B. Saunders.

[317] Zhang, K. 1990 Biting theorems for Jacobians and their applications, Anal. Nonlineare, 7, 345–366.

[318] Zhang, K. 1992 A construction of quasiconvex functions with linear growth at infinity, Ann. S. N. S. Pisa, 19, 313–326.

[319] Zhang, K. 1993 Rank-one connections and the three well problem, Trans. AMS (to appear).

Index